The Value of Signals
in Hidden Action Models

Contributions to Economics

www.springeronline.com/series/1262

Further volumes of this series can be found at our homepage.

Peter Meusburger/Heike Jöns (Eds.)
Transformations in Hungary
2001. ISBN 3-7908-1412-1

Claus Brand
Money Stock Control and Inflation Targeting in Germany
2001. ISBN 3-7908-1393-1

Erik Lüth
Private Intergenerational Transfers and Population Aging
2001. ISBN 3-7908-1402-4

Nicole Pohl
Mobility in Space and Time
2001. ISBN 3-7908-1380-X

Mario A. Maggioni
Clustering Dynamics and the Locations of High-Tech-Firms
2002. ISBN 3-7908-1431-8

Ludwig Schätzl/Javier Revilla Diez (Eds.)
Technological Change and Regional Development in Europe
2002. ISBN 3-7908-1460-1

Alberto Quadrio Curzio/Marco Fortis (Eds.)
Complexity and Industrial Clusters
2002. ISBN 3-7908-1471-7

Friedel Bolle/Marco Lehmann-Waffenschmidt (Eds.)
Surveys in Experimental Economics
2002. ISBN 3-7908-1472-5

Pablo Coto-Millán
General Equilibrium and Welfare
2002. ISBN 7908-1491-1

Wojciech W. Charemza/Krystyna Strzala (Eds.)
East European Transition and EU Enlargement
2002. ISBN 3-7908-1501-1

Natalja von Westernhagen
Systemic Transformation, Trade and Economic Growth
2002. ISBN 3-7908-1521-7

Josef Falkinger
A Theory of Employment in Firms
2002. ISBN 3-7908-1520-9

Engelbert Plassmann
Econometric Modelling of European Money Demand
2003. ISBN 3-7908-1522-5

Reginald Loyen/Erik Buyst/Greta Devos (Eds.)
Struggeling for Leadership: Antwerp-Rotterdam Port Competition between 1870–2000
2003. ISBN 3-7908-1524-1

Pablo Coto-Millán
Utility and Production, 2nd Edition
2003. ISBN 3-7908-1423-7

Emilio Colombo/John Driffill (Eds.)
The Role of Financial Markets in the Transition Process
2003. ISBN 3-7908-0004-X

Guido S. Merzoni
Strategic Delegation in Firms and in the Trade Union
2003. ISBN 3-7908-1432-6

Jan B. Kuné
On Global Aging
2003. ISBN 3-7908-0030-9

Sugata Marjit, Rajat Acharyya
International Trade, Wage Inequality and the Developing Economy
2003. ISBN 3-7908-0031-7

Francesco C. Billari/Alexia Prskawetz (Eds.)
Agent-Based Computational Demography
2003. ISBN 3-7908-1550-0

Georg Bol/Gholamreza Nakhaeizadeh/Svetlozar T. Rachev/Thomas Ridder/Karl-Heinz Vollmer (Eds.)
Credit Risk
2003. ISBN 3-7908-0054-6

Christian Müller
Money Demand in Europe
2003. ISBN 3-7908-0064-3

Cristina Nardi Spiller
The Dynamics of the Price Structure and the Business Cycle
2003. ISBN 3-7908-0063-5

Michael Bräuninger
Public Debt and Endogenous Growth
2003. ISBN 3-7908-0056-1

Brigitte Preissl/Laura Solimene
The Dynamics of Clusters and Innovation
2003. ISBN 3-7908-0077-5

Markus Gangl
Unemployment Dynamics in the United States and West Germany
2003. ISBN 3-7908-1533-0

Pablo Coto-Millán (Ed.)
Essays on Microeconomics and Industrial Organisation, 2nd Edition
2004. ISBN 3-7908-0104-6

Wendelin Schnedler

The Value of Signals in Hidden Action Models

Concepts, Application, and Empirical Evidence

With 11 Figures
and 9 Tables

Springer-Verlag Berlin Heidelberg GmbH

Series Editors
Werner A. Müller
Martina Bihn

Author

Wendelin Schnedler
University of Bristol
Department of Economics (CMPO)
12 Priory Road
Bristol BS8 1TN
United Kingdom
w.schnedler@bristol.ac.uk

ISSN 1431-1933
ISBN 978-3-7908-0173-6 ISBN 978-3-7908-2706-4 (eBook)
DOI 10.1007/978-3-7908-2706-4
Cataloging-in-Publication Data applied for
A catalog record for this book is available from the Library of Congress.
Bibliographic information published by Die Deutsche Bibliothek
Die Deutsche Bibliothek lists this publication in the Deutsche Nationalbibliografie; detailed bibliographic data is available in the Internet at <http://dnb.ddb.de>.

Softcover Design: Erich Kirchner, Heidelberg

SPIN 10982075 88/3130/DK-5 4 3 2 1 0 – Printed on acid-free and non-aging paper

Acknowledgements

I thank René Fahr, Anne Glatz, and Sher Verick for reading the manuscript and helping me improve it substantially. I am also indebted to my dissertation committee: Klaus Zimmermann (chair), Matthias Kräkel, and Jörg Oechssler. I benefited a lot from discussions with various researchers. Amongst them were: Paul Chen, Damon Clark, Bruno Crépon, Ingolf Dittmann, Tor Eriksson, Laurent Gobillon, Burhkhard Hehenkamp, Francis Kramarz, Joseph Lafranchi, Thierry Magnac, Gerd Mühlheusser, Stefan Napel, Georg Nöldeke, Jean-Marc Robin, Klaus Schürger, Urs Schweizer, Wolfgang Schwerdt, and Uwe Sunde. The following four institutions kindly provided me with stimulating working environments: Bonn Graduate School of Economics, Centre de la Recherche en Economie et Statistique in Paris, Institute for the Study of Labor in Bonn, and the Leverhulme Centre for Market and Public Organisation in Bristol. Finally, I am grateful to the following people for supporting me: Verónica Boiani (at least at some stage), Benedikt Glatz, Katja Haumann, Ellen Schnedler, Fabian Schnedler, and Henrik Schnedler.

Bristol, November 2003 *Wendelin Schnedler*

If thou wilt work, I have bread for thee; if thou wilt be honest, sober, and industrious, I have greater rewards to confer on thee – ease and independence.

Hector St. John De Crèvecoeur (1735-1813)
Letters from an American Farmer (1782)
Letter III: What is an American?

Contents

1 Introduction

Since the fundamental work of Walras (1874), markets have received particular attention by economists because they lead to an efficient allocation of goods and services. However, the proper functioning of markets rests on certain assumptions. For instance, the good or service which is to be traded must be clearly defined. This elementary requirement is often violated in reality, in particular when services are concerned.

Consider the example of railway workers who are hired to lay tracks. A labour contract which stipulates a fixed wage and defines the workers' task as "laying tracks" is rather unspecific. Workers may profit from this vagueness by reducing effort to a comfortable amount –as long as tracks are laid, they do not violate contract conditions. Thus, an imprecise definition of the service can result in inefficiently low efforts.

An obvious solution to this problem is a clearer definition of the service, but often this way is barred: To specify, for instance, all actions which are involved in laying tracks and which may vary with weather, surface and other conditions is far too complicated and too costly. Indeed, labour contracts seldom give a detailed account of the task of a worker. Alternatively to a more precise task description, the wage of the worker could be conditioned on information about the worker's performance. For example, the railway workers might be paid by the length of tracks laid so that they are motivated to exert more effort. Hence, if actions cannot be directly stipulated, it is still possible to indirectly induce more efficient actions by tying payments to signals, which convey information about these actions.

Therefore, the value of signals potentially lies in improving efficiency. But how large is this value? What determines it? And how much of it is contributed by which determinant? These and similar questions are examined in this monograph.

The value of signals unfolds within a mechanism which conditions payments on signals to indirectly induce an action; such a mechanism is called an *incentive scheme.* Designing an incentive scheme is problematic for at least two reasons. First, the signal may include a stochastic element, so that the wage of the worker becomes uncertain. For example, a storm might prevent the railway workers to advance with their regular pace, a shorter track length is laid, and the wage, which depends on track length, is reduced. On another day, weather conditions might allow a higher speed, and the wage is larger. If workers dislike this uncertainty in their income, they need to be compensated in form of a higher base wage. The trade-off between the aim to provide incentives and the preference of workers for a fixed income is well investigated (Holmström 1979, Shavell 1979, Grossman and Hart 1983). The second problem related to the design of an incentive scheme is effort misallocation. The benefit which the railway company draws from tracks is not only defined by their length. If tracks are laid in a hasty way, they wear down faster, and traffic must be interrupted to renew them. When workers are just paid by track length, they will tend to neglect this aspect. Their effort is not well-balanced between diligence and quantity; rather, it is distorted. Hence there also exists a trade-off between inciting high effort and allocating effort amongst different tasks. The role of effort misallocation has firstly been recognised by Holmström and Milgrom (1991) and then discussed by many authors (e.g. Feltham and Xie 1994, Prendergast 1999, Baker 2002). Nevertheless, effort misallocation has remained a vague concept which is usually described by means of examples or special models but not in form of a general definition.

This book provides a general definition of effort misallocation. In addition, the relationship between misallocation and the value of signals is theoretically described, applied to the example of firm performance measures, and empirically analysed.

The second chapter embeds the current work in the literature, defines misallocation, and relates it to the value of signals in incentive schemes. The necessity of designing an incentive scheme arises from the problem that actions which benefit an employer and are carried out by a worker cannot be contractually fixed. This places the analysis in the broader framework of non-contractibility. The reader is introduced to different models in which non-contractibility causes inefficiencies; various settings are considered in which inefficiencies are reduced by some mechanism. The standard model to describe actions which are not contractible is the *hidden action* or *moral hazard model.* A subsection of the second chapter is devoted to this model, which serves as the basis for later extensions. The worker in this model only has to decide about a single variable: the amount of effort which he exerts. Thus, the worker cannot misallocate effort between tasks, and the only disadvantage of using signals lies in the uncertainty it may impose on the worker's income. To provide a richer model, a multiple task version of the hidden action model is considered. In this setting, there might exist effort allocations which cannot be implemented by any incentive scheme; in other words, not all effort allocations may be feasible. In addition, implementing some actions might impose less uncertainty on the worker than implementing others. Both uncertainty and feasibility can distort the effort allocation between tasks.

To compare the distortionary effect of different signals, misallocation needs to be measured. One of the original contributions of this monograph is to provide a widely applicable and readily interpretable measure for the degree of misallocation: the costs of misallocation. These costs are defined by the willingness of both parties to pay for a signal which leads to the appropriate allocation of effort between tasks. This definition of misallocation is compared with measures proposed by Feltham and Xie (1994) and Baker (2002). The definition is also used to decompose the costs which arise from the inability to stipulate the action of the worker. This decomposition allows fruitful insights into the strengths and weaknesses of signals.

It is shown that misallocation and uncertainty considerations are to some degree interdependent in optimal incentive schemes. For example, the optimal mechanism might sacrifice the desired effort allocation to impose less uncertainty on the worker. However, misallocation can also occur independently from uncertainty. Both cases are illustrated by examples. Finally, the analogy between misallocation and uncertainty in incentive schemes and the concept of bias and variance in statistics is outlined.

The third chapter applies the proposed concept of misallocation costs to a specific class of signals: firm performance measures. This chapter shows how misallocation costs lead to a better understanding of the value of firm performance measures and how to derive empirical implications about the degree to which firm performance measures are used in incentive schemes.

To explain the use of firm performance information in incentive schemes, the advantages and disadvantages of the respective signals are discussed. Since firm performance measures are often affected by the actions of many workers, a potential disadvantage of them is free-riding: Workers might rely on coworkers to generate a benefit from which they also profit. Referring to Holmström (1982), it is argued that using signals which encompass the actions of multiple agents alone is not sufficient to generate free-riding. If only workers know their true costs of exerting effort, the case for free-riding seems to be stronger because workers may pretend to have high costs to justify lower effort exertion. But even this situation does not imply free-riding (McAfee and McMillan 1991). Therefore, free-riding has to be ruled out as a disadvantage of firm performance measures.

Some useful conclusions about firm performance measures are drawn from benefit signals. A model by Lazear (1995), which describes the utilisation of benefit signals, is examined. In contrast to this model, which explains the utilisation of benefit signals by selection effects, an alternative model is devised which is novel in the sense that it builds on the concept of misallocation costs and allows to view the utilisation of signals in a new light. The alternative model assumes that benefit signals reflect the benefit of worker's actions better but have a higher variation than effort signals. Hence, employing benefit signals reduces misallocation but imposes additional uncertainty on the worker. Thus it pays to use benefit signals whenever the reduction in misallocation costs exceeds the increase in the compensation for uncertainty. The model serves as an example how the concept of misallocation costs can be applied to generate new insights into the nature, value, and optimal utilisation of signals.

Like benefit signals, firm performance measures tend to reduce misallocation at the price of higher uncertainty, so that the following prediction is derived: Firm performance measures are expected to be more intensively used when misallocation is possible and damaging. To check this claim empirically, the intensity of firm performance measures and the costs of misallocation need to be operationalised. It is argued that firm performance pay can proxy the intensity of firm performance measures and that task descriptors indicate costs of misallocation.

The fourth chapter examines the empirical evidence for the relationship between firm performance measures and the costs of misallocation. Other empirical studies which analyse incentive provision are presented, and differences and similarities to the present study are pointed out. A thorough test of the relationship between firm performance pay and work characteristics requires data on individual jobs. The empirical analysis in this chapter is novel in the sense that it is the first analysis to use job level data to examine this relationship. The unique data set which provides this opportunity is the Enquête sur la Coût de la Main-d'Oeuvre et la Structure des Salaires (ECMOSS) of the French National Statistical Institute (INSEE). Unfortunately, a special feature of the question on payment supplements complicates matters substantially: Firm performance is only observed if it belongs to the five largest payment supplements and is thus censored below a varying threshold. The development of a consistent estimator for this situation is another major achievement presented in this monograph, because it can be utilised for a variety of censoring problems. Finally, estimations are carried out, results are presented, and the relationship between work descriptors, control variables, and firm performance pay is examined.

The final chapter summarises and discusses the major findings.

2

Effort Misallocation and the Value of Signals

What is the relationship between effort misallocation and the value of signals in hidden action models? In this class of models, relevant actions cannot be contractually fixed. The inability of one party to commit to the efficient action entails an efficiency loss. In other words, there are costs of non-contractibility. Signals may help to reduce the costs of non-contractibility: By conditioning payments on signals, otherwise not contractible actions can be induced.

But the induction of an action has its price. One aspect of this price is compensation for uncertainty: Conditioning payments on random signals burdens the worker with uncertainty for which the latter needs to be compensated, if he or she dislikes uncertainty. Every additional signal which carries information about the exerted action helps to reduce this uncertainty (if its informational content is not yet incorporated by another signal), leads to a lower compensation, and thus has a value. This is the essence of the famous *information principle* (Holmström 1979). However, the validity of the information principle is limited to the case where the worker only has to decide about the effort for a single task, so that the action of the worker comprises setting a single effort level. In this case, uncertainty induces two types of costs: (i) a reduced effort level because a full internalisation of the beneficial effects of the action would impose too much uncertainty on the worker, and (ii) a compensation paid to the worker for the uncertainty imposed on him or her. Additional information helps to reduce both costs.

If the action encompasses setting several effort levels because the worker has to carry out multiple tasks, a third aspect becomes important: the misallocation of effort. Hence, using a signal now has three effects: It may lead to higher effort levels, reduce the compensation for uncertainty, and induce the worker to re-allocate effort in some helpful or harmful way. The net effect determines the overall value of signals.

The significance of misallocation and its associated costs express themselves in many ways. First, there are multitasking models in which it is optimal not to use signals, so there must be a third type of costs besides effort reduction and compensation for uncertainty. Second, there is anecdotal evidence on real life incentive contracts that led to results which cannot be explained solely on the basis of effort reduction and compensation. Even though misallocation seems to be the key to many phenomena, its character has remained astonishingly vague: While these phenomena are attributed to effort misallocation, no explicit definition exists. Research is hampered by this vagueness. A rigorous and general definition of misallocation which allows a unified treatment of the various observations loosely related to misallocation is highly desirable. Such a definition is the innovative element of this chapter.

The inability of employer and worker to specify important aspects of their contractual relationship is not unique to the problem at hand. The interaction between employer and worker as well as the type of non-contractibility have been modelled in various ways. Section 2.1 provides a structured overview of the literature and includes an informal introduction to the single as well as the multiple effort hidden action model, which form the background for the present analysis.

In Section 2.2, the costs resulting from the inability to specify the action of the worker are computed for the multiple effort model. Then, misallocation costs are defined and the overall costs are decomposed into different components related to the feasibility of effort allocations and uncertainty. Finally, measuring misallocation by its costs is compared to a measure put forward by Baker (2002).

Effort misallocation and uncertainty are related. In fact, the desire to reduce uncertainty may lead to a distorted allocation of effort amongst tasks. But uncertainty is no requirement for misallocation to occur. Section 2.3 illustrates these statements by examples. It also explores the analogy of the trade-off between misallocation and uncertainty to the trade-off between bias and variance in statistics. The main results of this chapter are summarised at the end.

For illustrative purposes, most of the theoretical analysis is framed in terms of an employer-worker relationship. Nevertheless, definitions and results apply to other applications of hidden action model as well: For example, the framework can be employed to describe and analyse insurance markets where the insurer is identified with the employer and the policy holder with the worker.

2.1 Problems of Non-contractibility

Consider a situation in which two parties contemplate exchanging a good or a service, but in which an important characteristic of this good or service cannot be contractually specified. Take, for instance, a worker who either knows more about his or her productivity than the employer or can affect the employer's benefit. Depending on the nature of the characteristic and the type of interaction between the two parties, different modelling frameworks are at hand. Section 2.1.1 presents three respective classes of models. Section 2.1.2 deals with the particular case of non-contractible actions and describes mechanisms to overcome this deficiency. Section 2.1.3 informally introduces the reader to the static hidden action model, which forms the backbone of the present analysis. In Section 2.1.4, the hidden action model is extended from one-dimensional to multiple non-contractible actions, the *multitasking* model.

2.1.1 Different Types of Non-contractibility

This section introduces various approaches to model non-contractible characteristics in economic interactions and defines the approach that is pursued here.

Hidden Information, Signalling, and Statistical Discrimination

Akerlof (1970) has shown that markets can break down if the non-contractible characteristic is fixed and unknown to one party. In his model, the non-contractible characteristic is the quality of a good. A different example is the productivity of a worker which is unknown to the employer. The literature suggests two ways how to deal with this problem. In Spence's *signalling model* (1973), the worker undertakes an activity which credibly signals productivity, for example he or she obtains a degree. In the *hidden information* or *adverse selection model* which was first studied by Rothschild and Stiglitz (1976) and Wilson (1977), the unknown information is elicited by cleverly designed screening contracts; here, workers reveal their productivity by choosing different workloads.

The presence of unknown relevant characteristics usually entails an efficiency loss. Signalling and screening mechanisms diminish this loss. If additional information is available which is related to the unknown characteristic, the loss can be further reduced. In the context of an employer-worker relationship, the information could be a trait which is correlated with the worker's productivity. Phelps (1973) uses this idea to explain racism and sexism as the result of payoff maximisation by the employer. The employer treats workers with different traits differently. This happens not because of individual tastes, as argued by Becker (1957), but because traits contain information about productivity. Since the relationship between trait and productivity is stochastic, conditioning on the trait is to the disadvantage of those workers who are highly productive but happen to exhibit low productivity traits. Treating a group with a common trait equally although they individually differ in the characteristic of interest is known as *statistical discrimination*. Altonji and Blank (1999) give an overview on statistical discrimination, and Schnedler (2002) places statistical discrimination in the context of the hidden information model.

Hidden Action Models

In the discussion described above, the important and non-contractible characteristic was fixed and unobservable. But often this characteristic can be influenced by one of the parties after an agreement. For example, a worker who was hired by an employer might set his or her effort level without being controlled. Again, inefficiencies may arise from the non-contractibility. The involved parties might have conflicting interests; the worker could prefer low effort, while the employer desires high effort. A framework to study non-contractible characteristics which can be influenced by one of the parties is the *hidden action* or *moral hazard model*. Foundations to this modelling framework were laid by Holmström (1979), Shavell (1979), and Grossman and Hart (1983). Like the hidden information and the signalling model, it has since made its way into economic textbooks and is considered an important tool to analyse economic questions.

The term "hidden action model" is misleading in so far as the problem is not so much that the action is hidden but rather that the action cannot be enforced by a contract. If both parties observe the action but cannot credibly reveal their observations to a third party (e.g. a court), it is still impossible for the third party to verify whether the contract was violated by one of the two contracting parties. In essence, the enforcement problem remains.

Incomplete Contracts and the Hold-up Problem

Yet another stream of the literature deals with the case that the non-contractible characteristic is *ex ante* unknown to both parties and will eventually be revealed to both of them. The characteristic is important for the optimal choice of an action in the future, and it is observable but not verifiable. Because the characteristic is not verifiable, contracts cannot condition the action on the characteristic, and parties have to negotiate about the action after the characteristic is observed. Choosing the optimal action rather than any other action generates a surplus, and the two parties will bargain how to share this surplus. If both parties have some bargaining power, each will obtain at least part of the surplus. If, in addition, the size of the surplus depends on investments made before the characteristic is revealed, some of the gain of the investment accrues to the other party, the investor will not fully internalise the positive consequences of the investment, and the investment will be below the efficient level. This effect is known as the *hold-up problem*, and a respective model was put forward by Grossman and Hart (1986) to explain the existence and the boundaries of firms. The approach initiated a vital area of research (see among others: Hart and Moore 1990, Che and Hausch 1999). Tirole (1999) provides an overview, classifies, and reviews this literature, which operates under the label of *incomplete contracts*. The label itself is a misnomer in the sense that contracts are incomplete in many settings that do not belong to this literature, for example in hidden information or hidden action models.

As the title of this monograph suggests, the analysis pursued here focuses on the hidden action model. Hold-up problems will not appear, and hidden information models will be absent, with the exception of the third chapter, where unobservable fixed characteristics play a minor role. The following paragraphs will present various mechanisms to implement desired actions in hidden action models.

2.1.2 Mechanisms to Enforce Non-contractible Actions

When the action of the worker is not contractible, mechanisms can be devised to induce the worker to exert effort. This section discusses some of these mechanisms.

Punishing by Dismissal

Shapiro and Stiglitz (1984) argue that the threat of being dismissed by the employer induces the worker to exert effort. Because dismissal only works as a threat if it is associated with a utility loss, it is assumed that dismissed workers become unemployed and receive a lower income. Thus, unemployment becomes a necessary ingredient to motivate employed workers. Workers receive wages higher than the wage offered to the unemployed and this generates room for more efficient production. This *efficiency wage theory* has been criticised for excluding contracts in which the worker posts a bond *ex ante* which he or she loses once fired: Losing the bond represents the same threat as becoming unemployed, so unemployment would not be needed as a discipline device (Carmichael 1985). There has been some debate on whether workers have the required access to credit markets to finance such bonds (Shapiro and Stiglitz 1985) and whether bonds can be substituted by upward-sloping wage profiles (Akerlof and Katz 1989).

The efficiency wage model is vague about the details of the contractual environment. In particular, it is not clear, why it is possible to condition the status of the employment relationship on the worker's action, whereas it is not possible to condition payments on it. The latter hints at the fact that effort is non-contractible. Thus, the relationship between worker's behaviour and dismissal is beyond explicit contractual enforcement and other reasons must exist for the employer and worker to adhere to the incentive scheme.

Self-enforcing Contracts

Most observed incentive schemes do not contractually condition remuneration on effort. Nevertheless, remuneration is tied to effort in many informal ways, for example by promotion and dismissal. The theory of *implicit contracts* aims at shading light on informal agreements which structure the relationship between employer and worker.

Arrangements which are not contractually fixed cannot be enforced by courts or other external institutions. They must be self-enforcing. But why should employer and worker be able to commit to a self-chosen arrangement? An answer is provided within the realm of the *theory of repeated games*: When interactions are repeated infinitely often, the opportunity exists to punish deviations from the agreement in subsequent interactions. Fudenberg and Maskin (1986) show that in such a context a wide range of equilibrium strategies, i.e. agreements, becomes sustainable. The possibility to sustain equilibria in infinitely repeated games by using the power of future interactions is the essence of a result that is known as *Folk theorem*.

This idea has been applied to the employer-worker relationship by Bull (1987) in order to explain why implicit contracts exist. MacLeod and Malcomson (1989) take this idea further to fill the gap in the model of Shapiro and Stiglitz (1984). They develop a model in which only the employment status and payments are contractible and where employers promise to reward effort and keep this promise. Additionally, involuntary unemployment occurs when workers do not believe the employer's promise. Levin (1999) studies self-enforcing incentive schemes in an infinitely repeated setting where the information about effort is imperfect. He shows that in optimal contracts discretionary transfers are made to the worker any time when the performance signal exceeds a threshold.

Self-enforcing agreements need not be limited to two parties. Kandori (1992) extends the principle of the Folk theorem to multilateral agreements, which implies that groups can jointly sustain rules and provides a rationale for social norms. Applying Kandori's results to the employer-worker context explains why workers' choice of action is also guided by norms. Kandel and Lazear (1992), Encinosa III, Gaynor, and Rebitzer (1997), and Huck, Kübler, and Weibull (2001) take group norms as exogenously given and analyse their effect on effort provision. Baker, Gibbons, and Murphy (2002) examine the choice between self-enforcing and externally enforced contracts. In this model, the rate at which employer and worker discount the gains from future interactions and the payoff outside the relationship exogenously determine the optimal choice. Similar to the models on group norms, the self-enforcing mechanism is not explicitly modelled by Baker, Gibbons, and Murphy (2002). Che and Yoo (2001) go one step further and explain how self-enforcing and externally enforced mechanisms interact to sustain group effort between two workers in an infinitely repeated setting.

Self-enforcement and Preferences

To justify self-enforcing arrangements, the infinite repetition of the interaction between employer and worker is essential. If there is a known end to the interaction, for example, because there is a mandatory retirement, the party who carries out the last action has no incentive to stick to the joint arrangement. Thus, if the worker is supposed to exert high effort, he avoids the effort costs, and if the employer is supposed to reward the agent, she keeps the money for herself. As there is no future interaction, neither of them can be punished for the deviation. Hence, the arrangement is not self-enforcing shortly before retirement. The same argument applies to the moment just before the arrangement loses its self-enforcing character, so that the arrangement collapses altogether. The reason for the collapse is essentially the same as the reason for a breakdown of cooperation in a finitely repeated prisoner's dilemma.

In reality and in economic experiments, one frequently observes informal arrangements to be rather stable although interactions are known to come to an end. Kreps, Milgrom, Roberts, and Wilson (1982) develop a model which explains these observation by the presence of some economic agents which have preferences for adhering to a rule. Others are trying to imitate these agents to build up a reputation. Up to the last interaction, the result is observationally equivalent to a situation where everybody is obeying the rule.

The concept of Kreps, Milgrom, Roberts, and Wilson (1982), which was intended to explain cooperation in the prisoner's dilemma, was adapted by Brown, Falk, and Fehr (Forthcoming), who claim that the possibility to engage in repeated interactions can overcome contractual incompleteness problems in employer-worker relationships. They support this claim by experimental evidence and a theoretical model. In their theoretical model, they suppose that some workers have preferences for *reciprocity*, that is they reciprocate generous payments. Employers and workers can interrupt the relationship and search for a different partner after every interaction. Employers have a reason to pay above market wages even in the final interaction since some agents reciprocate generous payments by exerting high effort. This creates a rent to all agents in the final repetition. The prospect of receiving this rent instead of being dismissed gives agents an incentive in the last but one repetition if they believe in the generosity of their employer. To merit this trust of the agent, the employer has to be generous in this period as well. Thus, there will be a rent created for the agent in the last but one repetition. The same argument can now be applied iteratively until the first interaction. The experimental observations of Brown, Falk, and Fehr (2002) fully agree with their model.

Fehr and Schmidt (2000) find experimental evidence for the deliberate choice of less complete contracts by the employer. They trace this behaviour back to *fairness preferences*: A fair employer needs no expensive technology to verify the effort of the worker. She simply reciprocates effort exertion with a bonus payment. Workers expect to have a fair employer with some probability and thus exert effort. They do so even when they are selfish because they hope to be rewarded.

Fehr and Falk (1999) show in a simple market experiment that *reciprocity* suffices to establish a market wage above the minimum wage although workers outnumber employers. Employers are reluctant to accept low wage offers by workers because they fear that, once employed, these workers shirk. On the other hand, they accept workers which offer to work for high wages, because some of these reciprocate the increased wage with greater effort and thereby more than repay the costs of their salary. To some extent, this story can be regarded as the experimental counterpart of the efficiency wage theory.

Renegotiation in Hidden Action Models

Closely linked to self-enforcement is the issue of renegotiation. The possibility to renegotiate complicates self-enforcing agreements because the parties have an additional opportunity to deviate. But renegotiation also has effects when the incentive scheme is externally enforced. Fudenberg and Tirole (1990) and Ma (1994) allow for renegotiation of the contractual terms in the hidden action model after the action has been taken but before the signal is realised. Since the action is fixed at the moment of renegotiation, it is only the unsystematic noise component of the signal which is not yet determined. The separation from noise and action suggests that the insurance aspect can be disentangled from the incentive aspect. It turns out that this is not the case and that it is crucial whether the worker (Ma 1994) or the employer (Fudenberg and Tirole 1990) has the right to propose a new contract at the renegotiation stage. In the case of a proposing worker, the set of implementable actions is reduced, and it becomes impossible to induce costly actions with probability one, while renegotiation has neither a beneficial nor a harmful effect in the second case.

In a repeated hidden action model, future re-negotiations may influence the design of the optimal contract. Schnedler and Sunde (2002) examine a model in which the worker's action has an effect on his outside option in future re-negotiations with the employer. To have a better stance in this future renegotiation, the employer prefers retaining control over the worker by integrating him into her firm rather than providing incentives and outsourcing him.

Scope of the Present Analysis

Certain themes which were discussed in the previous paragraphs are absent from the analysis pursued here: There will be neither self-enforcement nor repetition or renegotiation.

Although self-enforcing arrangements play a major role to solve incompleteness problems in reality, only contracts which can be enforced by third parties are studied in the following. The reason is simple: The primary interest of this monograph is the analysis of the value of signals. Different enforcement methods require different features of the signal to be used in the incentive scheme. For instance, self-enforcing mechanisms can often rely on observable signals, while externally enforced mechanisms require contractibility of the signal. But apart from this distinction, the designer of the incentive scheme faces the same problems. Should an additional monitoring device be bought? Should a signal be used and if so, to what degree? Since these questions, which concern the value of signals, arise independent from the enforcement mechanism, it is justified to abstract from the method of enforcement. To avoid the more complex structure of self-enforcing arrangements, it is thus assumed without loss of generality that all signals are contractible. The action itself will of course remain non-contractible.

The present work is also limited to a simple one-shot hidden action problem without re-negotiations between contract acceptance and implementation of the contractual specifications. Certainly, repeated interactions and re-negotiations play an important role in real life. But the reduced framework employed here is sufficient to generate important and interesting insights into how signals are used.

Additional Literature

The value of signals in incentive schemes is related to several other topics. Incentives are often provided by promotions. Therefore, the study of careers and career paths has a lot of common ground with the study of incentive schemes. A prominent example from this field is the analysis of Baker, Gibbs, and Holmström (1994). A recent overview on this topic is provided by Gibbons and Waldman (1999). A particular form of incentive provision are individual employment contracts; the main insights in this field are summarised by Malcomson (1999). Explicit incentive schemes are often found for managerial positions. Hence, a lot of attention has been devoted to modelling executive compensation. For an example see Jensen and Murphy (1990), and Murphy (1999) for a review of the literature. The theoretical background for the study of incentive schemes is contract theory, which is described in numerous monographs (Macho-Stadler and Perez-Castrillo 1997, Salanié 1998, Schweizer 1999).

The value of signals does not only unfold in incentive schemes. Signals are for example also used when making decisions under uncertainty or otherwise incomplete information. So, the value of signals links up with the topic of uncertainty and information. Uncertainty and information forms a widely researched stream of economic literature which is partly summarised in the textbook of Hirshleifer and Riley (1992). This stream includes topics such as the value of delaying decisions until more information is available (Dixit and Pindyck 1994) or the pricing of options (see for example Gouriéroux and Jasiak 2001), to name just two.

Finally, the present work belongs into the realm of personnel economics, and a major source of reference in the following will be Lazear's monograph (1995) on this topic.

2.1.3 Optimal Incentives in the Single Effort Model

We build on a one-shot hidden action model in which contracts can explicitly condition on signals. This section introduces such a model following Holmström (1979) and Shavell (1979). Using the customary terminology, the party which carries out the action or effort should be called *agent*, while the other party, which benefits from the effort, is called *principal*. The former can be identified with the worker and will be referred to by male pronouns and the second can be identified with the employer and will be referred to by female pronouns.

If the agent exerts effort, he incurs costs $C(e)$. These costs are increasing in effort e. The principal obtains a benefit $B(e)$, which depends on the choice of the agent and increases in this effort either deterministically or in some stochastic way.

Because the benefit of the effort accrues to the principal, while the costs are shouldered by the agent, some mechanism is needed so that the agent internalises the beneficial effects of his effort on the principal. Without such a mechanism, the agent chooses the lowest possible effort level. To compensate the agent for his effort, the principal carries out a transfer τ. This transfer positively affects the *income of the agent*: $\iota = \tau - C(e)$. From this income, the agent draws a utility $u(\iota)$, where $u(\cdot)$ is a von-Neumann-Morgenstern utility function. The agent is assumed to be risk-averse, so that $u(\cdot)$ is concave, while the principal is supposed to be risk-neutral.

If the effort e were contractible, the transfer τ could be conditioned on e and the so-called first-best effort which maximises the difference between benefit $B(\cdot)$ and costs $C(\cdot)$ could be achieved. But neither e nor $B(\cdot)$ or $C(\cdot)$ can be verified before a court. Hence, it is not possible to condition τ on any of these quantities, and principal and agent must look for other means to link the transfer τ to the effort e; such a link will be referred to as an *incentive scheme*. It is assumed that a stochastic signal $S(e)$ is available which is verifiable and the distribution of which depends on the effort of the agent. The distribution is assumed to be common knowledge. Strictly speaking, and following the distinction of Knight (1921), the hidden action model is thus concerned with *uncertainty* and not with *risk*, because the latter refers to cases where the distribution is not common knowledge but assessed subjectively. Nevertheless, it is customary to speak of "risk" which is imposed on the agent instead of uncertainty. In addition, the agent is often called "risk-averse" when he dislikes uncertainty and "risk-neutral" when he is indifferent with respect to uncertainty; the terms "risk compensation" and "risk premium" are usually employed to refer to the compensation for uncertainty and "risk sharing" to describe a situation where both, principal and agent, shoulder some uncertainty. Since these terms are predominantly used in the literature, some of them are also utilised in the subsequent text. On the other hand, the term "risk" is avoided whenever it is possible without harming the readability.

Often, it is supposed that the signal is identical to the benefit of the principal $B(e)$. However, there is no reason to limit attention to signals based on the benefit *a priori*. Any signal related to the non-contractible effort is a candidate for an incentive scheme, and the term *performance measure* or *performance signal* is used here to refer to it. This choice of wording is justified by the fact that all these signals measure some aspect of the agent's performance. Indirectly, performance measures also carry information on the benefit since the benefit depends on effort. Signals which depend on effort only via the benefit function are called *benefit signals*.

The signal $S(e)$ allows to link the transfer τ to the effort e, but its stochastic character makes it an imperfect substitute for a deterministic link: The risk-averse agent will be exposed to uncertainty. In order for the relationship between principal and agent to be mutually beneficial, the agent needs to be indemnified for this uncertainty in addition to the compensation for the exerted effort. This has two well-known consequences which are, for example, described in Salanié (1998), Schweizer (1999) or any other standard textbook dealing with the hidden action model. First, the effort that can be induced by the optimal incentive scheme which is based on the signal $S(e)$ is lower than the first-best effort. Second, the agent shoulders some uncertainty and gets compensated. Thus, the incentive scheme using $S(e)$ generates a lower surplus than the first-best solution which can be attained when e is contractible. This loss induced by the impossibility of writing contracts in terms of e can be decomposed into two components: the compensation for uncertainty and the reduced benefit for the principal which stems from the decreased effort.

Holmström (1979) shows that this loss can be reduced if additional signals are employed in the optimal incentive scheme. This requires that an additional signal bears some information about e which is not yet reflected by the signals employed in the incentive scheme; that is, the signals already in use are not a sufficient statistic for e. This result has become known as the *information principle*.

This rather informal presentation of the hidden action model conceals substantial difficulties, which become apparent when the problem is analysed more formally. These difficulties are related to the properties of the signal $S(e)$. Even if $S(e)$ is deterministic, it must fulfil certain requirements so that it is possible to recover the effort e from observing the signal $S(e)$ and achieve the first-best solution. In particular, there must be a unique signal for the first-best effort level. This requirement is fulfilled if different signals are produced by distinct effort levels, that is if $S(e)$ is injective in e, which is, for example, the case if the signal strictly increases in e. If the signal $S(e)$ is stochastic, the construction of an optimal incentive scheme depends on some form of monotonicity as well: $S(e)$ must be stochastically larger for large values of e. The formulation *stochastically larger* implies a comparison of random variables. There are different ways to carry out this comparison: for example, based on expected values, probability distribution functions, or cumulative distribution functions of the random variables. The existence of an optimal incentive scheme rests on particular assumptions how the signal stochastically increases in effort.

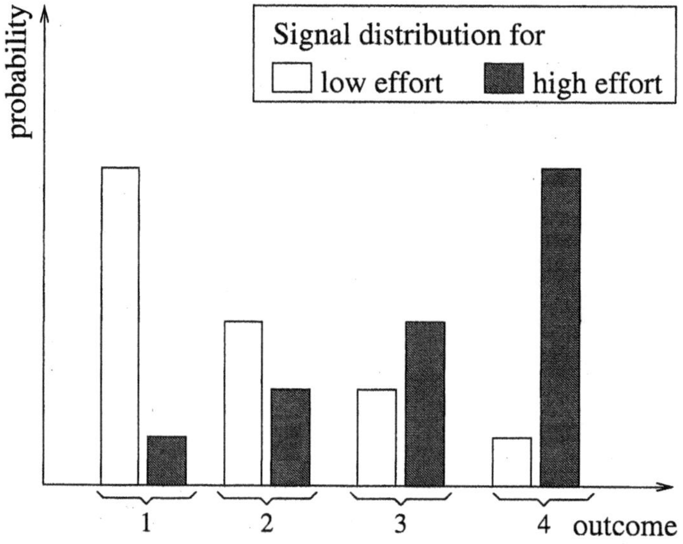

Fig. 2.1. Example for Monotone Likelihood Ratio Condition
Higher effort moves the probability mass to the right and increases the probability of high outcomes more than that of low outcomes.

A common assumption is based on the comparison of probability distributions. In the case of discrete effort and outcomes, it means that higher effort increases the probability of a high outcome at least as much as the probability of a low outcome, see Figure 2.1. This assumption is equivalent to the requirement that the ratio of the probability of a larger outcome versus a smaller outcome increases in effort. Since the quotient of probabilities can be seen as a likelihood ratio, it is known under the label *monotone likelihood ratio condition (MLRC)* (see Salanié 1998). In the case of two outcomes, this property is equivalent to first-order stochastic dominance, but generally it is a stronger concept (Müller and Stoyan 2002). Grossman and Hart (1983) propose the *convexity of the distribution function condition (CDFC)* as an alternative to the "Monotone Likelihood Ratio Condition". This condition is fulfilled if and only if the cumulative distribution function is convex in effort.

If benefit is stochastic and $S(e)$ is identical to it, $B(e) \equiv S(e)$, the debate on the properties of $S(e)$ can also be understood as requirements with regard to the benefit function. Schweizer (1999) defines a benefit function to be normal if its expected value is convex in effort; he also proves that this necessarily entails convexity of the cumulative distribution function in effort. Thus, the condition of Grossman and Hart (1983) follows from requiring a "normal" benefit function.

Finding conditions for a monotone stochastic relationship between effort and signal is not the only problem related to $S(e)$. The type of distribution also matters. If $S(e)$ follows a discrete distribution, the structure of the incentive scheme is manageable: It consists of a specification of transfers for a countable number of signals. Mathematically, the optimal incentive can be specified by countably many real numbers. If these numbers are taken from a compact space, the existence of an optimal incentive scheme is ensured. But if $S(e)$ is continuously distributed, the structure of the incentive scheme becomes very complex because the number of signals is not countable any more. Then, finding the optimal transfer scheme amounts to selecting a function, and the maximiser of the optimisation problem is a functional of the parameters. To guarantee the existence of a solution, the space from which the functions are taken must be compact. To avoid this issue altogether, one may restrict the class of functions and for example only allow for contracts which make linear use of signals. In some of the subsequent models, this shortcut is employed to analyse incentive schemes with continuous signals.

Another difficulty arises from the dimensionality of the effort. The incentive scheme is designed to induce the agent to exert a particular effort. Hence, it must fulfil the so-called incentive constraint; that is, the agent should have no incentive to deviate from the desired effort choice. Checking for the validity of this condition is relatively easy for a finite number of efforts: One simply has to compare the finite number of utility levels which the agent can reach with the respective effort choice. If the agent can choose amongst countably infinite or infinitely many effort levels, infinitely many comparisons have to be carried out. Then, checking or devising an incentive constraint becomes rather complicated, and it is common to resort to the *first-order approach.* This approach uses only the (necessary) first-order condition and assumes that if the agent has no incentive to deviate from the desired effort level by a margin, he also has no incentive to deviate by more than a margin. Mathematically, this requires that the first-order condition is necessary and sufficient.

2.1.4 Misallocation in the Multiple Effort Model

As long as effort is one-dimensional, it can only be too much or too little. If effort is costly, the latter is the case. Then, an incentive scheme can be used to increase effort and bring it closer to the first-best level. Observations suggest that one-dimensional effort models do not reflect relevant aspects of reality; the introduction of piece-rate schemes has frequently led to results that cannot be explained within these models.

A recent example was published in *The Economist* ("The Washington mayor's race – Watch out for Mr Blair" August 3rd, 2002, p. 38). Anthony Williams, the mayor of Washington D.C., needed to collect 2000 signatures of registered voters to get his name on the Democratic primary ballot in order to be eligible for re-election. He hired workers and paid them one dollar for each signature they obtained. The highly motivated workers gathered over 10,000 signatures. A closer inspection of the signatures revealed that global celebrities like Kofi Annan and Tony Blair were amongst the signers and are neither known to live in Washington D.C. nor to be eligible to vote; also, a surprising number of signers had the same hand-writing. All this raises doubts about the authenticity of the signatures. Apparently, the workers did what they were paid for (collecting signatures), but not what Anthony Williams wanted them to do (collect genuine signatures).

Something similar occurred when the American telephone company AT&T started paying programmers by the number of code lines. Effort was stimulated, and programmes became longer, but AT&T drew no benefit from these longer programmes (example taken from Prendergast 1999). Similarly, Lincoln Electric was able to increase the effort of secretaries by paying them by the number of keystrokes; some secretaries went even so far to use their lunch break hitting senselessly on one key (example taken from Baker 2002).

These examples illustrate that it is not sufficient to stimulate a particular effort when this effort only partly reflects the benefit of the principal. Anthony Williams needed authentic signatures to become eligible for re-election, but his incentive scheme only placed emphasis on the number of signatures not on authenticity. AT&T desired to improve their services, but longer programmes are a means to this end only if the quality of the programmes is not degrading. Lincoln Electric sought to enhance administration but got useless keystrokes.

In all three cases, workers responded to the incentive scheme with increased effort, but the increased effort along a particular dimension of the work had no positive consequences on the benefit since other dimensions were neglected: Effort allocation is distorted. This misallocation leads to costs which manifest themselves in a lower benefit for the principal: Mr. Williams suffers from a severe reputational loss, longer programs complicate maintenance, and keyboards wear down more quickly. In addition and more importantly, losses result from the foregone opportunity to use the paid labour in a more beneficial way.

To accommodate these phenomena in the hidden action framework, it needs to be extended to multiple efforts. Such an extension is the *multitasking model* by Holmström and Milgrom (1991). In a particular variant, the *home contractor model*, the optimal solution is a fixed wage. There is no incentive component, although an informative signal is available. This finding would have been unthinkable in the single effort case, as it contradicts Holmström's information principle (1979). However, with multidimensional efforts, a signal does not only have a uncertainty-reducing effect but also an allocative effect (Feltham and Xie 1994, Baker 2002, Schnedler 2003). Since a misallocation of effort induced by using a signal is costly, it is only optimal to use it when the benefit from uncertainty reduction outweighs the costs of misallocation. This idea will be vital for most of the following analysis.

One approach to multitasking problems is the re-assignment of tasks between workers such that the problem is minimised or disappears

(Holmström and Milgrom 1991, Hirao 1993, Itoh 1994). For example, Holmström and Milgrom (1991) point out that tasks which are difficult to monitor should be separated from tasks which are simple to monitor. However, they also admit that their model "oversimplifies these issues by assuming that there are no restrictions on how the principal may group tasks." In reality, the allocation of tasks seems to be less flexible because it is limited by technological requirements. For example, Mr. Williams cannot separate the collection of a signature from authenticity: It is impossible to hire a worker which transforms the signatures produced by another worker into authentic signatures. Since these limitations on task allocation between workers exist, it is important to analyse the design of incentive schemes under multitasking.

There exists an equivalent to Holmström's information principle (1979) for the multitasking case. It was devised by Feltham and Xie (1994) and can be applied if benefit, signalling technology, and contracts are linear. It states that an additional signal has no value if there exists a linear combination of the (old) signals which are already used, if this linear combination has the same expected value as the additional signal, and if the covariance between the linear combination and the old signals is equivalent to the covariance between the additional and the old signals. In the same restricted linear setting, Feltham and Xie (1994) as well as Baker (2002) develop measures for the allocative value of a signal which are discussed later in Section 2.2.5.

The multiple effort model contributes an important insight with respect to the use of information and its costs: It is not only the effect on the level of effort and on uncertainty compensation which determines the value of information but also the effect on the allocation of effort. Attempts to quantify allocative effects of signals have been sparse and limited to the class of linear benefit and signalling functions. The next section presents a more general approach to quantify allocative effects.

2.2 Decomposing Agency Costs

It is the aim of this section to identify different cost components which result from the non-contractibility of effort in a multitasking model. From the previous section it is clear that, besides the compensation for uncertainty paid to the agent and the reduced effort of the agent, there is potentially a third type of costs: the costs due to effort misallocation. The question what signals to use in an incentive scheme then boils down to a trade-off along three dimensions: risk sharing between principal and agent, incentives for the agent to exert effort, and incentives for the agent to allocate effort. As no general method existed to quantify the costs of misallocation of effort, this trade-off has long had only abstract character. Here, a definition for the costs of misallocation is suggested, so that the trade-off becomes concrete as the compensation for uncertainty, costs of reduced effort, and costs of misallocated effort can be computed and compared with each other.

Section 2.2.1 introduces the model and defines the overall loss from being unable to contract on effort. Generally, this loss can be attributed to uncertainty and feasibility. In Section 2.2.2, agency costs are decomposed into a component related to uncertainty and another component related to feasibility. In Section 2.2.3, uncertainty and feasibility costs are further broken down into an effort reduction and misallocation component, so that a detailed decomposition of the costs of noncontractibility becomes possible. In Section 2.2.4, this decomposition is used to disentangle the contribution of effort misallocation, effort reduction and uncertainty compensation. It is also used to compare the value of a collection of signals relative to another collection of signals. Finally, Section 2.2.5 compares misallocation costs with alternative measures of misallocation.

2.2.1 Agency Costs

In this section, the notation of the single effort model is taken up, formalised, and extended to describe a multitasking setting. Then, the costs of not being able to contract upon effort are derived. To do so, effort choice and surplus are determined for a context where effort is contractible (first-best). Subsequently, the optimal effort choice and the respective surplus are computed when effort is not contractible (second-best). The surplus difference are the *agency costs*; they indicate the maximal amount principal and agent are willing to spend on an agency which renders effort contractible.

The Model

Consider a multitasking hidden action problem in which an agent exerts a multidimensional effort $e \in I\!R^n$ for which he incurs costs $C(e) \in I\!R$, where $C(\cdot)$ increases and is continuous in all effort dimensions. A principal draws a benefit $B(e) \in I\!R$ from the effort, and this benefit increases in all effort dimensions. The principal can pay the agent a transfer τ. Without loss of generality, the benefit $B(e)$ is assumed to be deterministic in effort. If effort is contractible, the transfer τ can be conditioned directly on e. If effort is not contractible, it may only be conditioned on a signal random vector $S(e)$ whose distribution depends on e. The agent dislikes uncertainty more than the principal; to simplify matters, it is again assumed that the principal is risk-neutral while the agent is risk-averse and has a concave von-Neumann-Morgenstern utility function $u(\cdot)$. The utility for principal and agent when no effort is exerted is standardised to zero.

First-best Problem

Here, it is assumed that τ can be conditioned on e or a known non-stochastic and injective function of e, for example $B(\cdot)$ or $C(\cdot)$. In this case, contracts which specify e can be enforced by courts. Hence, the principal and the agent can write a contract that stipulates the effort which maximises the joint surplus, indemnifies the agent for his costs, and divides the surplus amongst them. Because the joint surplus is maximised, this contract defines the best outcome which principal and agent can ever achieve. The respective *first-best maximisation programme* is:

$$\text{argmax}_e B(e) - C(e). \tag{2.1}$$

Denote by E^{FB} the set of maximisers of this programme and by e^{FB} an element of this set. Suppose that the set is not empty; for example, this is the case if the surplus $\Psi(e) := B(e) - C(e)$ is concave and positive for some effort. Then, the *first-best surplus*, which is generated by the relationship between principal and agent, amounts to

$$\Psi^{\text{FB}} := B(e^{\text{FB}}) - C(e^{\text{FB}}), \quad \text{where } e^{\text{FB}} \in E^{\text{FB}}.$$

The same surplus can be achieved when the effort vector e (or a non-stochastic injective function of effort like benefit or effort costs) is observable to principal and agent: Principal and agent have to announce their observations and if these announcements do not match, they are severely punished. However, this supposes that principal and agent cannot renegotiate before the punishment is carried out. Finding themselves in a situation were they should be punished and are able to renegotiate, it is in the interest of both principal and agent to cancel the contract to avoid the punishment. Foreseeing this cancellation, neither party has an incentive to announce the true observation any more, and a true announcement cannot be part of a sequential equilibrium strategy. Thus, the mechanism of making observable effort contractible by using Draconian punishments when reports about the effort differ seems to be more of theoretical than of practical interest.

Second-best Problem

Now it is assumed that the transfer τ cannot directly be linked to the effort vector e. In this case, an indirect link between τ and the effort e is established by relating τ to the stochastic signal vector, which in itself depends on the effort e: The incentive scheme $\Upsilon(\cdot)$ assigns a transfer to every realisation s of the signal vector $S(e)$.

What is the best mechanism that principal and agent can devise? The mechanism should generate the highest possible joint surplus as in the first-best case, but it must also respect individually rational behaviour of both parties. To guarantee that the principal acts individually rational, she might be assigned the role of the mechanism designer. Individual rationality of the agent is then represented in the form of an *incentive constraint.* Additionally, it has to be ensured that the principal is not generating benefit by a mere redistribution from the agent to her. This is guaranteed by a *participation constraint.*

Having assigned the principal the role of the mechanism designer, she becomes the recipient of the surplus which is generated by the mechanism. This is just a convenient way of thinking about the problem and has no implications about the actual division of the surplus. Any such division could be achieved by an appropriate choice of the outside option of the agent, which influences the agent's share via the participation constraint. How the surplus is shared belongs to the realm of bilateral bargaining theory. The reader interested in this topic may, for example, consult Napel (2002). Here, the interest lies entirely in the construction of a mechanism which optimally uses contractible information in order to maximise the *joint* surplus of the relationship between principal and agent when effort is not contractible. Being the mechanism designer, the principal is made the advocate of the common cause; whatever benefits her is beneficial to both.

To maximise surplus, the principal has to solve the following programme, which only differs from the typical programme in textbooks such as Fudenberg and Tirole (1991), Salanié (1998), or Schweizer (1999) by the multidimensionality of effort and the signal:

$$\begin{aligned} & max_{\Upsilon(\cdot)} \; \mathrm{E}\left[B(e) - \Upsilon(S(e))\right] \\ \text{such that} \;\; & e \in \operatorname{argmax}_{\tilde{e}} \mathrm{E}\left[u\left(\Upsilon(S(\tilde{e})) - C(\tilde{e})\right)\right] \\ \text{and} \;\; & \mathrm{E}\left[u\left(\Upsilon(S(\tilde{e})) - C(\tilde{e})\right)\right] \geq 0. \end{aligned} \tag{2.2}$$

The second line represents the incentive constraint and the last line the participation constraint, and where $\mathrm{E}[\cdot]$ denotes the expected value operator.

The maximisation programme (2.2) explains why a deterministic benefit function can be assumed without loss of generality: The risk-neutral principal only cares about the expected benefit, and the risk-averse agent's payoff does not directly depend on the benefit. The latter is no restriction as any desired dependency between agent's payoff and benefit can be modelled using the signal vector $S(\cdot)$. For example, one component of this vector could be a noisy benefit measure.

Agency Costs when the Second-best Does Not Exist

It was already pointed out in Section 2.1.3 that the existence of a solution to the maximisation programme is not ensured in general. There are two reasons why a solution does not exist. Either it is impossible to design an incentive scheme which allows the principal and the agent to generate a surplus. Then principal and agent are left with their outside options, which were standardised to zero and consequently the surplus which is lost from being unable to contract on the action amounts exactly to the first-best surplus Ψ^{FB}. Or there exists an incentive scheme Υ' for any incentive scheme Υ such that Υ' generates a higher surplus. However, there must also exist a smallest upper bound on the surplus,$\overline{\Psi^{\mathrm{SB}}}$, which is smaller or equal to the surplus generated in the first-best setting. Then, the loss resulting from the inability to contract on the action can be taken to be the difference between the first-best surplus and the smallest upper bound: $\Psi^{\mathrm{FB}} - \overline{\Psi^{\mathrm{SB}}}$. For the remainder of the analysis, we assume that the second-best problem has a solution.

Agency Costs when the Second-best Exists

Let $\Upsilon^{\mathrm{SB}}(\cdot)$ be an incentive scheme which maximises the programme (2.2), E^{SB} be the set of corresponding effort choices, and e^{SB} be a particular optimal effort choice. Then the surplus generated by the second-best incentive scheme is:

$$\Psi^{\mathrm{SB}} := \mathrm{E}\left[B(e^{\mathrm{SB}}) - \Upsilon^{\mathrm{SB}}(S(e^{\mathrm{SB}}))\right].$$

Being unable to contractually fix effort amounts to the difference between the first-best and the second-best surplus. Thus, *non-contractibility costs (agency costs)* are defined to be equal to that difference:

$$C^{\mathrm{A}} := \Psi^{\mathrm{FB}} - \Psi^{\mathrm{SB}}. \tag{2.3}$$

The surplus loss represented by the agency costs is composed of various components. By decomposing the agency costs, it is possible to identify which aspects of the second-best incentive scheme contributed most to the loss and thus to assess the strength and weaknesses of incentive schemes. In particular, it can be identified whether the incentive scheme leads to effort misallocation. Hence, the decomposition of agency costs is one of the central issues in the following analysis.

Certainly, it is not necessarily the case that all components of the effort vector are either contractible or not contractible. Intermediate cases may occur in which some components of the effort vector are contractible, while others are not. For such intermediate cases, the optimisation programme slightly differs from (2.2): Effort components which are contractible are directly chosen by the principal whereas effort components which are not contractible are indirectly induced by the choice of the incentive scheme Υ. Since the choice of contractible efforts poses no problem and does not contribute to the costs of non-contractibility, the intermediate case is not further examined here.

2.2.2 Feasibility and Uncertainty Costs

We identify two sources of non-contractibility costs: feasibility and uncertainty. Due to feasibility, certain effort allocations cannot be implemented, while uncertain renders the implementation more expensive. In this section, we split agency-costs according to these two sources. First, feasibility, feasibility costs of effort, and the compensation for uncertainty are defined. These concepts help us to capture and structure the effects of non-contractibility on effort. Then, these effects are studied using a simplified second-best programme. Finally, the costs due to feasibility and uncertainty are computed and it is shown that their sum equals the non-contractibility costs.

Feasibility Costs of Effort

Grossman and Hart (1983) propose a two-step method to find an optimal incentive scheme (a scheme which solves the problem (2.2)). First, take any effort vector e and find the least cost incentive scheme Υ_e which implements it. Then, select the effort vector which generates the highest surplus. The existence of a least cost incentive scheme Υ_e, however, is not guaranteed for all efforts e. Those efforts which are feasible are collected in the *feasibility set*: $E^{\mathrm{F}} := \{e|\Upsilon_e \text{ exists}\}$. The *feasibility costs of effort* $C^{\mathrm{F}}(e)$ are defined to be prohibitively large for efforts which do not belong to that set:

$$C^{\mathrm{F}}(e) := \begin{cases} 0 \text{ for } e \in E^{\mathrm{F}} \\ \infty \text{ for } e \notin E^{\mathrm{F}} . \end{cases} \tag{2.4}$$

The feasibility costs of effort are a form to express the restriction which feasibility imposes on second-best effort. This restriction entails costs which will later form one of the components of non-contractibility costs.

Uncertainty Compensation

For every feasible effort, one can compute the difference between the expected transfer and the actual labour costs. This difference indemnifies the agent for the uncertainty imposed on him. Hence, the *compensation for uncertainty* $C^{\mathrm{U}}(e)$ is defined as:

$$C^{\mathrm{U}}(e) := \mathrm{E}\left[\Upsilon_e(S(e))\right] - C(e). \tag{2.5}$$

Alternatively, the terms *uncertainty compensation* and *uncertainty premium* are used to denote these costs. In the literature, it is sometimes referred to as *risk premium.*

The uncertainty compensation varies in the effort, because the implementation of different effort vectors might necessitate different incentive schemes, which in turn impose different uncertainty on the agent. If the agent is risk-neutral, uncertainty poses no problem, and the uncertainty compensation is zero. If the agent is risk-averse, it does not suffice to compensate the agent for the labour costs $C(e)$; in expectation, the agent needs to receive more than $C(e)$ to be willing to participate. Thus, the compensation for uncertainty is larger than zero.

Agency Costs of Effort

We already encountered agency costs, which described the costs of being unable to contract on effort. But what are the costs to implement a specific effort? In addition to labour costs $C(e)$, implementing effort e necessitates a surcharge composed of feasibility costs of effort $C^{\mathrm{F}}(e)$

and uncertainty compensation $C^{\mathrm{U}}(e)$. We define the *agency costs of effort* e to be:

$$C^{\mathrm{A}}(e) := C^{\mathrm{U}}(e) + C^{\mathrm{F}}(e).$$

This sum reflects the price that needs to be paid for getting a particular effort e when effort itself is not contractible.

Simplified Second-best Programme

We are interested in non-contractibility because it affects the effort choice. But why and how does it affect this choice? Consider the following maximisation programme, which is equivalent to the the second-best programme (2.2) in the sense that it has the same maxima and maximisers:

$$max_e \quad \underbrace{B(e) - C(e)}_{\text{real surplus}} \quad - \underbrace{\left(C^{\mathrm{U}}(e) + C^{\mathrm{F}}(e)\right)}_{\text{agency costs of effort}}. \tag{2.6}$$

This maximisation programme is equivalent to the second-best programme because it can be regarded as the second stage problem of the Grossman-Hart approach: The optimal incentive scheme for every effort choice Υ_e has already been computed on the first-stage respecting incentive and participation constraint, and now the effort vector which maximises the net benefit needs to be determined on the second stage.

Comparing (2.6) with the first-best problem (2.1) clarifies how the agency costs of effort affect the surplus and the effort choice: While the feasibility costs of effort $C^{\mathrm{F}}(e)$ limit the domain for effort, the uncertainty compensation $C^{\mathrm{U}}(e)$ alters its costs. The second-best efforts can be implemented by the second-best incentive scheme. Because second-best efforts are feasible, their feasibility costs are zero. This implies that the second-best surplus can be expressed as:

$$\Psi^{\mathrm{SB}} = B(e^{\mathrm{SB}}) - C(e^{\mathrm{SB}}) - C^{\mathrm{U}}(e^{\mathrm{SB}}).$$

The surplus difference between the first-best problem (2.1) and problem (2.6) has two sources: On the one hand, there are *direct costs*. These consist of the compensation for uncertainty $C^{\mathrm{U}}(e^{\mathrm{SB}})$, which needs to be paid to the agent when implementing the second-best effort. On the other hand, there are *indirect costs*. They arise from a difference in the effort level and allocation between the first-best and second-best efforts: $\Psi(e^{\mathrm{FB}}) - \Psi(e^{\mathrm{SB}}) + C^{\mathrm{U}}(e^{\mathrm{SB}})$.

Ultimately, it is our aim to identify how misallocation contributes to the indirect costs. First, however, the agency costs, which comprise both direct and indirect costs, will be attributed partly to uncertainty and partly to feasibility.

Uncertainty Costs

Which part of the agency costs are due to uncertainty? To identify the costs of uncertainty, first consider what happens if the incentive scheme operates in an environment without uncertainty. Such an environment can be simulated by suppressing uncertainty compensation. This defines the *certainty maximisation programme:*

$$max_e B(e) - C(e) - C^{\mathrm{F}}(e). \tag{2.7}$$

The maximisers or this problem, denoted by $e^{\mathrm{C}} \in E^{\mathrm{C}}$, serve as a reference point for the effort allocation if the agent is not risk-averse or if the signal is not random. The respective *certainty surplus* is:

$$\Psi^{\mathrm{C}} := B(e^{\mathrm{C}}) - C(e^{\mathrm{C}}) - C^{\mathrm{F}}(e^{\mathrm{C}}).$$

The difference between this surplus and the second best surplus Ψ^{SB} is only due to uncertainty and is called *uncertainty costs*:

$$C^{\mathrm{U}} := \Psi^{\mathrm{C}} - \Psi^{\mathrm{SB}}. \tag{2.8}$$

In contrast to uncertainty compensation, uncertainty costs are independent of the effort; they denote the overall costs when effort is chosen to minimise uncertainty costs. Uncertainty costs are always larger or equal to zero, since the second-best problem differs from the certainty problem by an additional restriction. When the uncertainty compensation for the second-best effort is calculated and subtracted from uncertainty costs, the result are the *indirect costs of uncertainty*: $\Psi^{\mathrm{C}} - \Psi^{\mathrm{SB}} - C^{\mathrm{U}}(e^{\mathrm{SB}})$. As shown later, these costs are composed of two ingredients: the reduction and the misallocation of effort due to uncertainty.

Feasibility Costs

Which impact does feasibility have on agency costs? Feasibility leads to costs because effort is limited to the set of feasible allocations E^{F}, which we represented in the maximisation programme by the feasibility costs of effort $C^{\mathrm{F}}(e)$. If desirable allocations are not feasible, a loss occurs.

In the previous section, we examined a certain environment. The only difference between this environment and the first-best situation are the the feasibility costs of effort. Thus, the surplus difference between the first-best, and the certainty setting is entirely caused by feasibility and *feasibility costs* are defined as

$$C^{\mathrm{F}} := \Psi^{\mathrm{FB}} - \Psi^{\mathrm{C}}. \tag{2.9}$$

Feasibility costs measure the overall effect of restrictions on the domain of effort choices. Hence, they are independent from effort and should not be confused with the feasibility costs of effort, which describe whether a particular effort is feasible or not.

If there are no restrictions on feasibility, that is if feasibility costs of effort are zero for all efforts, the optimal certainty effort choice coincides with the first-best effort choice. Then, the surplus of the certainty programme Ψ^{C} is identical to the first-best surplus Ψ^{FB}, the feasibility costs are zero, and agency costs are purely due to uncertainty.

Decomposing Agency Costs

It follows directly from the definitions of feasibility and uncertainty costs that agency costs can be decomposed in the following way:

$$C^{\mathrm{A}} \overset{(2.3)}{=} \Psi^{\mathrm{FB}} - \Psi^{\mathrm{SB}} = \Psi^{\mathrm{FB}} - \Psi^{\mathrm{C}} + \Psi^{\mathrm{C}} - \Psi^{\mathrm{SB}} \overset{(2.9) \text{ and } (2.8)}{=} C^{\mathrm{F}} + C^{\mathrm{U}}. \tag{2.10}$$

This implies that agency costs may be attributed to feasibility and uncertainty in a meaningful way. However, this decomposition does not tell us which part of non-contractibility costs is accounted for by misallocation.

2.2.3 Identifying Misallocation Costs

Misallocation of effort can result from two sources. First, it may be voluntarily chosen in order to reduce uncertainty. Second, the desired effort allocation might not be feasible. Accordingly, there are two types of misallocation costs: uncertainty-induced and feasibility-induced misallocation costs. But how can we assess these costs.

Towards a Definition of Misallocation Costs

Misallocation costs arise because the allocation of effort is not as it should be. More precisely, they can be seen as the difference between the surplus arising from a balanced in contrast to misallocated effort. The latter is, for example, the second-best effort. But how do we obtain a balanced effort? How can we get a reference with which to compare the second-best surplus? First-best efforts are balanced but the first-best surplus is a bad reference. Using the first-best surplus, all agency costs are attributed to misallocation. This is clearly not sensible because some of the difference between first-best and second-best cannot be attributed to misallocation; for example, this difference can be larger than zero in single-effort models, even though misallocation is not possible when there is only one effort. Both constituents of the loss due to non-contractibility (feasibility costs and uncertainty costs), are incurred partly because effort is misallocated but also because the overall effort level is reduced in comparison to the first-best. To find a sound definition of misallocation, we have to disentangle reduction from reallocation of effort.

To approach the issue, consider the following thought experiment. Suppose the principal operates an incentive scheme using a particular signal vector. Now, imagine that the principal is offered a monitoring device which allows her to elicit any effort allocation she wants as long as the labour costs do not exceed second-best levels. At the same time, the device ensures that the participation of the agent is fulfilled for the price of participation in the second-best programme. The principal is willing to spend at least as much as the uncertainty compensation for this device because it ensures participation. But most likely, the principal has a much higher willingness to pay because she values the freedom to choose an effort allocation which leads to a higher benefit. The device limits the overall effort in cost terms but not the allocation of efforts between tasks. So, it allows to disentangle reduction from reallocation; when using it, the effort is reduced but not reallocated. If we subtract the uncertainty premium from the principal's willingness to pay for the device, we can assess how much the freedom to allocate effort is worth to her. This quantity represents also the loss of being unable to get a balanced allocation. In other words, it captures misallocation costs.

Balanced Second-best Effort Allocation

In this section, we formalise the previous idea of assessing misallocation costs by deriving a balanced allocation and comparing its surplus with the second-best surplus. This poses a problem in so far as neither the balanced nor the second-best effort allocation have to be unique. Ultimately, it will thus be necessary to compare sets of efforts, but for now it should suffice to consider a particular second-best allocation e^{SB}.

To implement the second-best effort allocation e^{SB}, the principal has to incur the costs:

$$\mathcal{C}(e^{\mathrm{SB}}) = \underbrace{C(e^{\mathrm{SB}})}_{\text{labour costs}} + \underbrace{C^{\mathrm{U}}(e^{\mathrm{SB}}) + C^{\mathrm{F}}(e^{\mathrm{SB}})}_{\text{agency costs of effort}},$$

where the feasibility costs of effort are zero, $C^{\mathrm{F}}(e^{\mathrm{SB}}) = 0$, because the existence of an optimal incentive scheme which induces the effort choice e^{SB} implies that this effort is feasible.

Now, suppose a situation which is slightly different from the second-best. In this situation, the principal is granted a budget which amounts to the costs of implementing the second-best effort allocation $\mathcal{C}(e^{\mathrm{SB}})$. She is forced to pay the same money on agency costs of effort, but she is free to use the remainder $C(e^{\mathrm{SB}})$ in any way she wants. In particular, there are no restrictions imposed on the effort allocation between tasks. The maximisation programme defined by these assumptions is the following:

$$max_e B(e) - C(e) - C^{\mathrm{U}}(e^{\mathrm{SB}}) \text{ such that } C(e) \leq C(e^{\mathrm{SB}}). \qquad (2.11)$$

The constraint ensures that the principal is not spending more on labour than in the second-best case. Overall, the expenditure maximally amounts to the same level as in the second-best case: $\mathcal{C}(e^{\mathrm{SB}})$. Differing from the second-best case, any misallocation of the effort is eliminated because agency costs of effort are fixed in effort. Consequently, any maximiser of (2.11) represents a second-best effort allocation which is balanced and we denote it by $e^{\mathrm{BSB}} \in E^{\mathrm{BSB}}$.

Another intuition why effort allocation in the maximisation problem (2.11) is undisturbed is the following: The effort choice depends only on real terms, namely the actual benefit and the actual costs. In addition, problem (2.11) differs from the first-best problem (2.1), which is certainly balanced, only by the budgetary restriction: $C(e) \leq C(e^{\mathrm{SB}})$. This restriction limits overall effort but not the allocation between different efforts. Thus, the effort allocation resulting from the maximisation programme (2.11) is called *balanced second-best effort allocation*, while the maximisation problem is called *balanced second-best programme*. The surplus generated by the balanced programme is:

$$\Psi^{\mathrm{BSB}} := B(e^{\mathrm{BSB}}) - C(e^{\mathrm{BSB}}). \tag{2.12}$$

Binding Budget Constraint

Notice that any remainder of the budget designated for real labour costs is lost. Hence, the principal has an interest to spend the whole budget when faced with the balanced second-best programme. To see formally why it is optimal to use up the whole budget, suppose there exists an effort choice e which maximises (2.11) such that $C(e) < C(e^{\mathrm{SB}})$. As an alternative to this effort choice, consider e' which is identical to e in all but one arbitrary chosen component which is slightly larger in e' than in e. Due to the continuity of $C(\cdot)$, the alternative choice e' is not violating the budget constraint: $C(e') \leq C(e^{\mathrm{SB}})$. In addition, the benefit of the alternative choice e' is larger because the benefit increases in effort: $B(e') > B(e)$. Overall, an effort choice e with $C(e) < C(e^{\mathrm{SB}})$ cannot be optimal, and continuity ensures that the constraint holds with equality for the maximiser e^{BSB}: $C(e^{\mathrm{BSB}}) = C(e^{\mathrm{SB}})$. This implies that balanced efforts lie in a hyper-plane: The principal is free to chose $n-1$ efforts and the n-th effort is then determined automatically. It simply uses up whatever is left of the budget.

When effort is one-dimensional, the constraint implies that the effort must be equal to the second-best effort. Intuitively, the principal does not gain from the freedom to allocate between efforts because there is only one effort. Hence, she cannot do better than picking the second-best effort allocation.

Uncertainty-induced Misallocation Costs

To define the *uncertainty-induced misallocation costs*, take any balanced second-best effort allocation $e^{\mathrm{BSB}} \in E^{\mathrm{BSB}}$ and compare the surplus it generates with the surplus which is generated by a second-best allocation $e^{\mathrm{SB}} \in E^{\mathrm{SB}}$:

$$C^{\mathrm{UM}}(e^{\mathrm{SB}}) := \Psi^{\mathrm{BSB}} - \Psi^{\mathrm{SB}}. \tag{2.13}$$

Breaking down surpluses into benefits and labour costs and using the fact that the budget constraint is binding allows us to show that the uncertainty-induced misallocation costs are the difference between the balanced benefit and the second-best benefit:

$$C^{\mathrm{UM}}(e^{\mathrm{SB}}) = B(e^{\mathrm{BSB}}) - B(e^{\mathrm{SB}}).$$

In other words, the value of avoiding misallocation is the additional benefit the principal is able to generate when she is not confined to the second-best effort allocation. Conversely, the loss of having to respect the implicit restrictions on allocation imposed by the compensation for uncertainty yield the uncertainty-induced misallocation costs.

The uncertainty-induced misallocation costs are always larger than or equal to zero because the balanced programme is less restrictive than the second-best programme; it grants additional freedom with respect to effort allocation. Formally, it is always possible to choose the second-best effort allocation e^{SB} in the balanced second-best programme (2.11) and hence we can at least obtain the second-best surplus Ψ^{SB}.

It has already been shown for the one-dimensional case that balanced and second-best effort allocation coincide. According to the definition in (2.13), this implies that misallocation costs are zero. Thus, the definition respects the elementary consistency requirement that no misallocation costs should occur if effort is not misallocated.

The definition of uncertainty-induced misallocation costs relies on the second-best implementation costs $\mathcal{C}(e^{\mathrm{SB}})$. In particular, this implies that misallocation costs depend on e^{SB}. If there is a unique solution to the second-best programme (2.6), then the cardinality of E^{SB} is one and misallocation costs are unique as well. If there are multiple maximisers, neither the implementation costs $\mathcal{C}(e^{\mathrm{SB}})$ nor misallocation costs need to be unique. This unpleasant feature is inevitable: If we want to make a statement about misallocation, we have to relate the balanced effort allocation to some distorted reference allocation. If this reference is not unique, there cannot be a unique measure of misallocation. Given the set of maximisers, it is, however, possible to determine a lower and an upper bound for misallocation costs: $\underline{C}^{\mathrm{UM}} := \inf_{e^{\mathrm{SB}} \in E^{\mathrm{SB}}} C^{\mathrm{UM}}(e^{\mathrm{SB}})$ and $\overline{C}^{\mathrm{UM}} := \sup_{e^{\mathrm{SB}} \in E^{\mathrm{SB}}} C^{\mathrm{UM}}(e^{\mathrm{SB}})$.

The multiplicity of misallocation costs might be due to very different reasons: Perhaps some second-best incentive schemes do not induce a unique effort choice, or there are multiple second-best incentive schemes, and each induces a different effort. The former case has knife-edge character because neither principal nor agent have any preference over distinct effort choices; in fact, the indifference goes so far that the risk-averse agent is perfectly willing to randomise between efforts. This is no proof for anything, but it hints at the implausibility of this case to occur; indeed, this situation arises in none of the later examples.

In the second case, where the difference in optimal efforts is due to variation in incentive schemes, multiplicity of misallocation costs is not a curse but a blessing. It allows to discriminate between distinct incentive schemes which generate the same surplus. To find the *uncertainty-induced misallocation costs of an incentive scheme*, one simply has to plug the effort vector induced by this incentive scheme into (2.13).

Uncertainty-induced Reduction Costs

Finally, there are costs which are due to reduced effort. Like in a single effort moral hazard model, effort is reduced because keeping it at the first-best level would impose too much uncertainty on the agent. This lowers the surplus and hence uncertainty leads to costs of effort reduction. How large is this reduction? We have to compare the surplus which principal and agent achieve in a certain environment with some other surplus. As we are trying to identify a mere reduction and not a misallocation of effort at this point, a natural and neutral reference is the surplus of the balanced second best. Consequently, the *uncertainty-induced reduction costs* are defined as

$$C^{\mathrm{UR}}(e^{\mathrm{SB}}) := B(e^{\mathrm{C}}) - C(e^{\mathrm{C}}) - \left[B(e^{\mathrm{BSB}}) - C(e^{\mathrm{BSB}})\right]. \tag{2.14}$$

Like misallocation costs, reduction costs depend on the choice of the second-best effort e^{SB}. This may not be obvious from formula (2.14), but the dependency is hidden in the balanced effort e^{BSB}, which is constructed on the basis of the second-best effort as depicted in (2.11).

Decomposing Uncertainty Costs

Now, uncertainty costs can be decomposed. First, express the uncertainty costs in terms of benefit and costs from the balanced, second-best, and certainty programme:

$$\begin{aligned} C^{\mathrm{U}} = \Psi^{\mathrm{C}} - \Psi^{\mathrm{SB}} = B(e^{\mathrm{C}}) - C(e^{\mathrm{C}}) \\ - B(e^{\mathrm{SB}}) + C(e^{\mathrm{SB}}) + C^{\mathrm{U}}(e^{\mathrm{SB}}) \end{aligned}$$

Then, recall that $C(e^{\mathrm{BSB}}) = C(e^{\mathrm{SB}})$ because the full budget for actual labour costs is spent. A respective replacement, adding and subtracting $B(e^{\mathrm{BSB}})$, and rearranging terms leads to:

$$\begin{aligned} C^{\mathrm{U}} = \Psi^{\mathrm{C}} - \Psi^{\mathrm{SB}} = {} & B(e^{\mathrm{BSB}}) - B(e^{\mathrm{SB}}) \\ & + B(e^{\mathrm{C}}) - C(e^{\mathrm{C}}) - \left[B(e^{\mathrm{BSB}}) - C(e^{\mathrm{BSB}})\right] \\ & + C^{\mathrm{U}}(e^{\mathrm{SB}}). \end{aligned}$$

Applying the definitions of misallocation and reduction costs finally yields a complete decomposition of uncertainty costs into the different cost types:

$$C^{\mathrm{U}} = \Psi^{\mathrm{C}} - \Psi^{\mathrm{SB}} = \underbrace{C^{\mathrm{U}}(e^{\mathrm{SB}})}_{\text{uncertainty compensation}} + \underbrace{\underbrace{C^{\mathrm{UR}}(e^{\mathrm{SB}})}_{\text{reduction}} + \underbrace{C^{\mathrm{UM}}(e^{\mathrm{SB}})}_{\text{misallocation}}}_{\text{indirect uncertainty costs}} \tag{2.15}$$

All components have a natural interpretation. Uncertainty leads to a lower overall effort level as expressed by the reduction term, to a misallocation of effort towards tasks which are better measured, and to compensation for the risk-averse agent.

We now know, how the contributing factors to uncertainty costs can be disentangled using the idea of balanced second-best efforts. But uncertainty is not the only factor leading to agency costs. What about feasibility? Is it possible to decompose feasibility in a similar fashion?

Feasibility-induced Misallocation Costs

In the preceding paragraph, uncertainty costs were decomposed into effort reduction costs, effort misallocation costs, and the uncertainty premium. Now, the same is done with feasibility costs. Like indirect uncertainty costs, feasibility costs arise from two sources: effort reduction and effort misallocation. Being unable to achieve certain effort allocations, the principal has to respond by demanding other effort allocations. If the principal is not able to get what she desires, it might be optimal to ask for less, so that there is effort reduction.

The misallocation and reduction induced by the feasibility of effort can be separated perfectly analogous to the decomposition of indirect uncertainty costs. The only difference is that we now start from the certainty programme instead of the second-best programme and eliminate the effect of the feasibility costs of effort instead of the uncertainty compensation to get a balanced effort allocation.

The labour cost incurred by the worker for the certainty effort are $C(e^{\mathrm{C}})$. The employer is now allowed to spend this sum without having to respect the restrictions imposed by feasibility:

$$max_e B(e) - C(e) \text{ such that } C(e) \leq C(e^{\mathrm{C}}). \quad (2.16)$$

The resulting effort choice $e^{\mathrm{BC}} \in E^{\mathrm{BC}}$ can be considered as *balanced certainty effort*; the respective surplus is Ψ^{BC}. The *feasibility-induced misallocation costs* are defined as the difference between the surplus which is generated by the balanced effort and the distorted certainty effort:

$$C^{\mathrm{FM}}(e^{\mathrm{C}}) := \Psi^{\mathrm{BC}} - \Psi^{\mathrm{C}}.$$

Again, the principal spends the whole labour cost budget on effort so that the constraint binds: $C(e^{\mathrm{BC}}) = C(e^{\mathrm{C}})$.Like uncertainty-induced misallocation costs, feasibility-induced misallocation costs can thus be expressed as a reduction of benefit:

$$C^{\mathrm{FM}}(e^{\mathrm{C}}) = B(e^{\mathrm{BC}}) - B(e^{\mathrm{C}}).$$

Hence, feasibility-induced misallocation costs are associated with the benefit loss due to the misallocation of effort.

Feasibility-induced Reduction Costs

If feasibility leads to lower effort, this can be measured by the surplus difference between the first-best and the balanced certainty programme. Thus, *feasibility-induced reduction costs* are defined as:

$$C^{\mathrm{FR}}(e^{\mathrm{C}}) := \Psi^{\mathrm{FB}} - \Psi^{\mathrm{BC}}.$$

The issue of uniqueness has already been discussed for the uncertainty-induced costs; the problem and the respective discussion also apply to feasibility-induced costs and are not repeated here.

Decomposing Feasibility Costs

The costs of feasibility C^{F} can now be decomposed into an effort reduction and an effort allocation component:

$$C^{\mathrm{F}} = C^{\mathrm{FR}}(e^{\mathrm{C}}) + C^{\mathrm{FM}}(e^{\mathrm{C}}).$$

Since we identified the components of feasibility costs, it is possible to decompose agency costs into various components including effort misallocation and effort reduction costs.

Detailed Decomposition of Agency Costs

Non-contractibility costs have been decomposed into feasibility and uncertainty costs, which in turn comprise misallocation and reduction costs. By combining the two steps, one obtains a detailed decomposition of non-contractibility costs which includes misallocation costs. For the complete decomposition, write out and expand agency costs:

$$\begin{aligned} C^{\mathrm{A}} = \Psi^{\mathrm{FB}} - \Psi^{\mathrm{SB}} &= \underbrace{B(e^{\mathrm{FB}}) - C(e^{\mathrm{FB}}) - \left[B(e^{\mathrm{BC}}) - C(e^{\mathrm{BC}})\right]}_{=C^{\mathrm{FR}}(e^{\mathrm{C}})} \\ &+ \underbrace{\left[B(e^{\mathrm{BC}}) - C(e^{\mathrm{BC}})\right] - \left[B(e^{\mathrm{C}}) - C(e^{\mathrm{C}})\right]}_{=C^{\mathrm{FM}}(e^{\mathrm{C}})} \\ &+ \underbrace{\left[B(e^{\mathrm{C}}) - C(e^{\mathrm{C}})\right] - \left[B(e^{\mathrm{BSB}}) - C(e^{\mathrm{SB}})\right]}_{=C^{\mathrm{UR}}(e^{\mathrm{SB}})} \\ &+ \underbrace{B(e^{\mathrm{BSB}}) - B(e^{\mathrm{SB}})}_{=C^{\mathrm{UM}}(e^{\mathrm{SB}})} + C^{\mathrm{U}}(e^{\mathrm{SB}}). \end{aligned} \tag{2.17}$$

Re-ordering terms and using the definitions of the cost components which are summarised in Table 2.1, we get:

$$\begin{aligned} C^{\mathrm{A}} = \Psi^{\mathrm{FB}} - \Psi^{\mathrm{SB}} &= \underbrace{C^{\mathrm{UR}}(e^{\mathrm{SB}}) + C^{\mathrm{FR}}(e^{\mathrm{C}})}_{\text{reduction of effort}} \\ &+ \underbrace{C^{\mathrm{UM}}(e^{\mathrm{SB}}) + C^{\mathrm{FM}}(e^{\mathrm{C}})}_{\text{misallocation of effort}} \\ &+ \underbrace{C^{\mathrm{U}}(e^{\mathrm{SB}}).}_{\text{uncertainty compensation}} \end{aligned} \tag{2.18}$$

This representation of the agency costs exhibits its different sources: They arise from reduction of effort, misallocation of effort, and the compensation for uncertainty. Misallocation and reduction can be voluntarily incurred to save on uncertainty compensation, or they might be forced by the non-feasibility of desired allocations. However, the main interest of this book does not concern agency costs, but the value of signals. The next section shows how the decomposition of costs can be employed to assess different aspects of the value of incentive schemes and signals.

Table 2.1. Cost components of non-contractibility

	Indirect costs		Direct costs
	Effort		
	Reduction	Misallocation	
Uncertainty	$C^{\mathrm{UR}}(e^{\mathrm{SB}})$	$C^{\mathrm{UM}}(e^{\mathrm{SB}})$	$C^{\mathrm{U}}(e^{\mathrm{SB}})$
Feasibility	$C^{\mathrm{FR}}(e^{\mathrm{C}})$	$C^{\mathrm{FM}}(e^{\mathrm{C}})$	

2.2.4 The Value of Signals

It seems natural to identify the value of a signal vector $S(e)$ with the surplus which it generates in an incentive scheme. Here, we follow this approach and use the previous results to decompose the value into components related to misallocation, reduction, and direct uncertainty costs. We get a detailed inventory of the different aspects which determine the value of using a signal vector. The signals, which are represented by a signal vector, may also be compared to other signals. The surplus difference informs us which signals are more apt for the construction of an incentive scheme. By decomposing the surplus difference, we find out why the signals are more useful.

Decomposition of the Value of Signals

Second-best effort and surplus depend on the signal vector $S(\cdot)$ which is employed in the incentive scheme. Up to now, there has only been one signal vector and it was not necessary to express this dependence in the notation. Subsequently, there will be more than one signal vector. To clarify that the signal vector $S(\cdot)$ is used by the second-best incentive scheme Υ^{SB}, led to a particular second-best effort e^{SB}, generated the second-best surplus Ψ^{SB}, entails feasibility costs C^{F}, and so forth, a subindex is added to the respective notation: Υ_S^{SB}, e_S^{SB}, Ψ_S^{SB}, C_S^{F}, etc. With the new notation a concise definition of the value of a signal becomes possible. The *value of a signal vector* $S(\cdot)$ is given by the second-best surplus which is generated when it is used:

$$V(S) := \Psi_S^{\mathrm{SB}}.$$

Solving for the second-best surplus in the decomposed agency costs (2.18), one obtains a decomposition of the value of the signal vector S:

$$\begin{aligned} V(S) = \Psi_S^{\mathrm{SB}} &= \Psi^{\mathrm{FB}} - C^{\mathrm{U}}(e_S^{\mathrm{SB}}) \\ &- C_S^{\mathrm{FR}}(e_S^{\mathrm{C}}) - C_S^{\mathrm{FM}}(e_S^{\mathrm{C}}) \\ &- C_S^{\mathrm{UR}}(e_S^{\mathrm{SB}}) \ - C^{\mathrm{UM}}(e_S^{\mathrm{SB}}). \end{aligned} \tag{2.19}$$

This expression depicts how the various cost terms reduce the first-best surplus to yield the second-best surplus which can be generated using the signal S: Limited feasibility leads to effort reduction and effort misallocation; uncertainty considerations have a similar effect and, in addition, entail uncertainty compensation costs. Thus, the inner structure of the effect of signal $S(\cdot)$ is unveiled; the sources of the loss can be identified and related to each other, so that the strengths and weaknesses of the signal become apparent.

Biasing and Balanced Signals

Are misallocation costs really informative about the quality of a signal? To check this, we define the allocative quality of a signal independent of misallocation costs: A signal vector is called *biasing*, if there is another signal vector which leads to a higher benefit at the same labour costs and uncertainty compensation. If a signal vector is not biasing, we say it is *balancing*. Indeed, it turns out that these definitions are closely related to misallocation costs. If signals are balancing, there is no room to increase the benefit by using other signals with the same uncertainty compensation. Thus, the effort allocation is balanced which implies that misallocation costs are zero. Conversely, it is true that misallocation costs of zero indicate that there is no scope to improve on the effort allocation. This implies that there cannot be a signal which leads to a better effort allocation. Overall, signals are balancing if and only if their misallocation costs are zero. Hence, misallocation costs are indeed describing the allocative quality of a signal.

Decomposition of the Relative Value of Signals

It is also possible to compare two different signal vectors $\tilde{S}(\cdot)$ and $S(\cdot)$. By computing the difference of the value of both signal vectors, the better signal vector can be identified. Moreover, the reason for the superiority can be uncovered with the help of the decomposition (2.19):

$$\begin{aligned} V(\tilde{S}) - V(S) = \Psi_{\tilde{S}}^{\mathrm{SB}} - \Psi_{S}^{\mathrm{SB}} = {} & C_{S}^{\mathrm{FR}}(e_{S}^{\mathrm{C}}) - C_{\tilde{S}}^{\mathrm{FR}}(e_{\tilde{S}}^{\mathrm{C}}) \\ & + C_{S}^{\mathrm{FM}}(e_{S}^{\mathrm{C}}) - C_{\tilde{S}}^{\mathrm{FM}}(e_{\tilde{S}}^{\mathrm{C}}) \\ & + C_{S}^{\mathrm{UR}}(e_{S}^{\mathrm{SB}}) - C_{\tilde{S}}^{\mathrm{UR}}(e_{\tilde{S}}^{\mathrm{SB}}) \\ & + C_{S}^{\mathrm{UM}}(e_{S}^{\mathrm{SB}}) - C_{\tilde{S}}^{\mathrm{UM}}(e_{\tilde{S}}^{\mathrm{SB}}) \\ & + C_{S}^{\mathrm{U}}(e_{S}^{\mathrm{SB}}) - C_{\tilde{S}}^{\mathrm{U}}(e_{\tilde{S}}^{\mathrm{SB}}). \end{aligned} \quad (2.20)$$

This decomposition allows to assess the comparative advantage of $\tilde{S}$ in relation to S along various dimensions rather than simply stating that one signal generates a higher surplus. It might even be the case that $S(\cdot)$ is superior to $\tilde{S}(\cdot)$ with respect to some dimensions, while the opposite is true for other dimensions. Knowing this, conclusions about the utilisation of $\tilde{S}(\cdot)$ relative to $S(\cdot)$ can be drawn which cannot be derived by considering only the overall difference of the value of information. Examples for respective conclusions are provided in the third chapter.

In the setting of Holmström's (1979) information principle and in the context of Feltham and Xie (1994), the signal vectors only differ by a component which is included in vector $\tilde{S}(\cdot)$, but not in $S(\cdot)$. These articles discuss whether the additional component leads to a higher value of $\tilde{S}(\cdot)$ in relation to $S(\cdot)$. The potential of the additional component to increase surplus lies solely in the reduction of uncertainty compensation. Decomposition (2.20) shows that the relative value of an additional signal can also be discussed in terms of other costs. This also explains why Holmström's information principle (1979) does not hold in the multitasking model: A smaller uncertainty compensation does not make a signal valuable if it is – for example – outweighed by feasibility-induced misallocation costs.

2.2.5 Alternative Measures of Misallocation

Although the idea of misallocation has been prevalent in the literature since Holmström and Milgrom's seminal contribution (1991), there have been few attempts to quantify it. They are confined to the linear exponential normal model (LEN). This section presents this model, two approaches to measure misallocation in this model, and discusses them together with misallocation costs.

The Linear Exponential Normal Model

The linear exponential normal model is a relatively popular setting for hidden action models as it allows for simple closed-form solutions. The term linear exponential normal model (LEN) was coined by Spremann (1987); the model was extended to multitasking with linear signals by Feltham and Xie (1994). It is their setting which should be described here. In this setting, the costs of exerting effort along n dimensions are assumed to be quadratic and separable $C(e) = \sum_{i=1}^{n} \frac{e_i^2}{2}$ and the benefit is a linear function of efforts:

$$B(e) = (\boldsymbol{\delta})'e = (\delta_1, \ldots, \delta_n) \begin{pmatrix} e_1 \\ \vdots \\ e_n \end{pmatrix},$$

so that $\boldsymbol{\delta}$ is a coefficient vector which denotes the marginal benefit resulting from each effort component.

The agent exhibits *constant absolute risk-aversion (CARA)*; his utility function is $u(\iota) = -e^{-r\iota}$, where r is the *Arrow-Pratt measure of risk-aversion* and ι is the income of the agent which is the difference of transfers $\Upsilon(\cdot)$, which he receives, and costs $C(\cdot)$, which he incurs: $\iota := \Upsilon(\cdot) - C(\cdot)$.

Effort in itself is not contractible but there is a set of performance signals $\{S_i\}_{i=1,\ldots,m}$ which are linear functions of the efforts disturbed by a normally distributed noise term with mean zero and variance σ_i^2:

$$S_i(e) = (\mathbf{d^i})'e + \epsilon = (d_1^i, \ldots, d_n^i) \begin{pmatrix} e_1 \\ \vdots \\ e_n \end{pmatrix} + \epsilon_i.$$

The coefficient vector $\mathbf{d^i}$ indicates the marginal effect of effort on the expected value of the signal. This expected value is a linear combination of efforts: $\mathrm{E}[S_i(e)] = \mathbf{d^i}e$. A special case occurs when the expected signal and the benefit coincide. Because this implies that the signal and benefit vector are congruent ($\mathbf{d^i} = \boldsymbol{\delta}$), we use the term *congruent* signal. Since a congruent signal is a function of the benefit, it is also a benefit signal.

In the multitasking LEN-model, signals are completely characterised by the coefficient vector $\mathbf{d^i}$ and the variance σ_i^2. But there are good reasons to represent them differently to better isolate the effect of uncertainty. Variance alone does not capture uncertainty because a signal gets less noisy, if the informative part in form of the length of the coefficient vector $\mathbf{d^i}$ increases relative to the variance. In other words, it is the signal-noise ratio $\rho_i = \frac{\mathbf{d^i}'\mathbf{d^i}}{\sigma_i^2}$ which determines the uncertainty imposed on the agent and not the variance σ_i^2. The effect of uncertainty is the same for two signals, if the signal-noise ratio stays constant. Holding the length of the coefficient vector constant, there is a unique relationship between variance and signal-noise ratio. Thus, it is also possible to standardise the length of the coefficient vector and characterise signals by coefficient vector and signal-noise ratio.

Contracts in the LEN-model are restricted to be linear functions of the signals. Mirrlees (1999) has proven that linear functions are not optimal under the conditions of the LEN-model, so the restriction is not innocuous. However, linear functions are convenient to compute and to interpret and Holmström and Milgrom (1987) provide some justification, how they arise as optimal functions in a slightly more involved setting, where the principal cannot control when the agent carries out the effort. The transfer from the principal to the agent is:

$$\Upsilon(s) = w_0 + \sum_{i=1}^{m} w_i s_i,$$

where w_0 is the *base wage*, w_i are the *performance wage rates*, and $w_i s$ is the *performance pay*. The special form of the agent's utility function and the normality of the error term leads to a simple expression of the certainty equivalent of the agents expected payoff (see, for example, Kräkel 1999 or Salanié 1998). This *certainty equivalent* is:

$$CE(e) = \mathrm{E}\left[u(\iota(\tilde{S}, e))\right] = \iota(\mathrm{E}(\tilde{S}), e) - \frac{r}{2}\sum_{i=1}^{m} w_i^2 \sigma_i^2 = \iota(S, e) - \frac{r}{2}\sum_{i=1}^{m} w_i^2 \sigma_i^2,$$

where $\iota(s, e) = \Upsilon(s) - C(e)$ is the income of the agent.

As the outside option of the agent is standardised to zero, the agent accepts a base wage which exactly offsets the certainty equivalent and the principal chooses: $w_0 = -\sum_{i=1}^{m} w_i s_i + \sum_{i=1}^{m} w_i^2 \sigma_i^2 + C(e)$. This implies that we do not have to care about the base wage subsequently.

If the marginal benefit of efforts is different from their marginal effect on the signal, the signal represents the benefit imperfectly. This suggests that differences between $\boldsymbol{\delta}$ and $\mathbf{d^i}$ translate to a misallocation of effort and that a definition of misallocation based on differences in the two coefficient vectors is sensible. But how should the distance between the vectors be measured?

Feltham and Xie's Measure

Feltham and Xie (1994) use the fact that the ratio of second-best efforts in this particular linear setting equals the ratio of the respective signal coefficients:

$$\frac{e_1^{\mathrm{SB}}}{e_2^{\mathrm{SB}}} = \frac{d_1^i}{d_2^i}.$$

Further, the ratio of first-best efforts equals the ratio of the benefit coefficients:

$$\frac{e_1^{\mathrm{FB}}}{e_2^{\mathrm{FB}}} = \frac{\delta_1}{\delta_2}.$$

Thus, the first and second-best effort ratio are equal if the signal coefficient vector coincides with the benefit coefficient vector. The following term is zero when this is the case:

$$(\delta_1 d_2^i - \delta_2 d_1^i)^2.$$

Accordingly, Feltham and Xie (1994) define this term as the *measure of non-congruity.*

Ultimately, the misallocation induced by a signal is related to its use in an incentive scheme. If the coefficient vector $\mathbf{d^i}$ and the respective error term is multiplied by a fixed number z, the effect of this change can be revoked by dividing the signal realisation by z before using it in an incentive scheme. Hence, the multiplication with z should not alter the measured congruity of the signal. Indeed, the willingness to pay for a signal does not change when the signal is multiplied with z. Respectively, misallocation costs resulting from the signal do not alter. Feltham and Xie's measure of non-congruity, however, increases by the factor z^2. Consequently, their measure assigns different congruity to signals which are perfectly equivalent when used in an incentive scheme.

Baker's Measure

Graphically, multiplying the coefficients by a number changes the length of the coefficient vector while the direction remains unaltered. The angle between the signal vector $\mathbf{d^i}$ and the benefit vector δ stays the same if the signal vector is multiplied with a positive number. Moreover, the angle also represents congruity between signal and benefit and can thus potentially be used to characterise whether a signal is biasing or balancing.

If the angle between signal and benefit coefficient is zero, coefficient vectors point into the same direction and the expected signal represents the benefit. If the angle is orthogonal, the expected signal has no relation to the benefit and cannot be used to incite effort. When the angle is 2π, the signal vector points again in the same direction as the benefit vector. There is a periodicity of 2π in the relationship between expected signal and benefit.

Baker (2002) proposes the cosine of the angle as a measure of congruity. The cosine has the advantage that it respects the periodicity, leads to values between zero and one, and takes on the largest value if benefit and signal coefficients point into the same direction and the lowest value if the signal is orthogonal. Alternatively, Baker could have used the sine to obtain a measure of congruity rather than misallocation.

Although congruity seems to be related to misallocation of effort, it is finally the surplus which counts. But the surplus of two signals pointing in exactly opposing directions is equal in the LEN-model: Rewarding an agent for good performance has the same effect as not punishing the agent when the performance was not bad. Formally, a signal can be multiplied with minus one; this changes the direction of the coefficient vector and the change will be revoked by the wage performance rates. The surplus is thus the same for vectors which are π apart. In other words, there is a periodicity of π with respect to the effect of the angle on the surplus. To take care of this periodicity, one could improve Baker's measure by using the square or absolute value of the cosine.

Baker's clever approach to use the angle between benefit and signal coefficient vector to determine misallocation, is founded on the idea that congruent signals are leading to the correct allocation of effort. But congruity only says something about the relation between the expected value of the signal and the benefit and it remains to be proven that congruent signals are always balancing. In fact, changing a seemingly harmless assumption in the LEN-model leads to congruent but biasing signals, so that congruity is not a good measure of misallocation.

Congruent but Biasing Signals

The following example shows that identical coefficients in a linear benefit function and a linear signal do not lead to the maximal payoff. Rather, this payoff is achieved if the coefficients if the signal deviate from the coefficients in the benefit function. This implies that a signal which is congruent leads to an effort allocation which can be improved upon; the efforts induced by identical coefficients in the benefit function and the signal are misallocated. This misallocation is not identified by

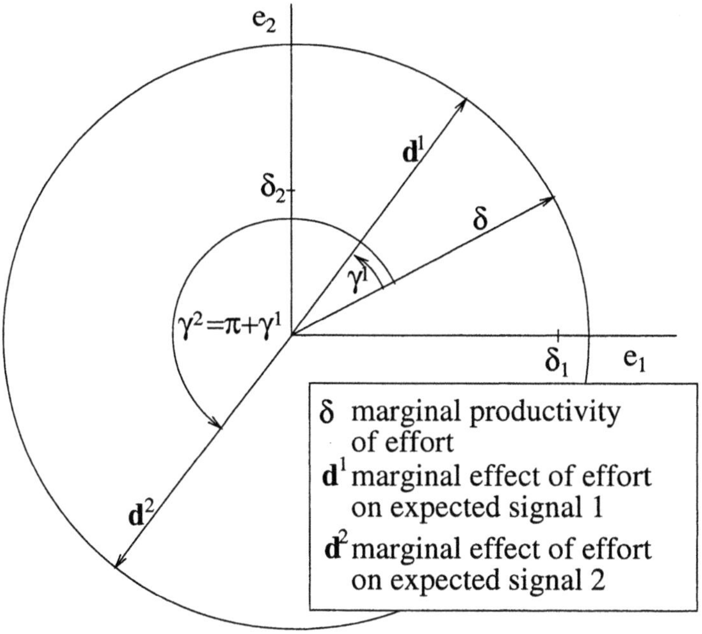

Fig. 2.2. Baker's measure of congruity
The expected signal and the benefit are represented by their coefficient vectors $\mathbf{d^i}$ *and* $\boldsymbol{\delta}$. *The cosine of the angle between benefit and signal vector measures congruity. Signals* $\mathbf{d^1}$ *and* $\mathbf{d^2}$ *represent the benefit equally well.*

the congruity measure of Baker (2002).

Consider the case where effort can be exerted along two dimensions: $e = (e_1, e_2)'$. The importance of each effort for the benefit is determined by a parameter δ. The parameter δ is taken to measure the marginal effect of the first effort on the benefit. Let us standardise the benefit vector to length one, then the marginal effect of the second effort is $\sqrt{1-\delta^2}$ and the complete benefit vector reads $\boldsymbol{\delta} = (\delta, \sqrt{1-\delta^2})'$, so that the benefit can be expressed as:

$$B(e) = \delta e_1 + \sqrt{1-\delta^2}\, e_2, \ \delta \in [0,1].$$

Likewise, d measures the marginal effect of the first effort on the expected signal. We also standardise the signal vector to length one, so that the marginal effect of the second effort on the signal is $\sqrt{1-d^2}$:

$$\tilde{S}(e) := S(e) + \epsilon := de_1 + \sqrt{1-d^2}\, e_2 + \epsilon, \text{ with } \epsilon \sim I\!N(0, \sigma^2).$$

Because the vector length is standardised, the signal-noise ratio is $\rho_i = \frac{1}{\sigma_i^2}$ and the effect of uncertainty is constant whenever ρ_i or σ_i^2 is held fixed.

With respect to costs, we slightly deviate from the introduced model by allowing that one effort may be relatively cheaper than the other effort: $C(e) = \frac{ce_1^2+e_2^2}{2}$, where c is a real-valued positive cost parameter. If $c = 1$ both efforts have exactly the same costs (this is the case examined previously). For $c > 1$, the first effort is more expensive. For $c < 1$ the agent prefers exerting the second effort.

Using the first order conditions of the concave surplus $B(e) - C(e)$, we get the first-best effort allocation $e_1^{\mathrm{FB}} = \frac{\delta}{c}$ and $e_2^{\mathrm{FB}} = \sqrt{1-\delta^2}$. This allocation leads to the first-best surplus: $\Psi^{\mathrm{FB}} = \frac{1}{2}\left(\frac{\delta^2}{c} + 1 - \delta^2\right)$. In the second-best case, the agent is compensated for the incurred uncertainty; the base wage w_0 is set so that he is just willing to participate and the principal incurs the respective uncertainty compensation costs: $C^{\mathrm{U}}(e) = \frac{r}{2}w_1^2\sigma^2$. The agent chooses the effort which maximises his expected payoff; because only the wage rate and the labour costs depend on effort, this choice maximises also the difference between the expected performance pay and the labour costs:

$$\mathrm{E}\left[w_1\tilde{S}(e)\right] - C(e) = w_1\left(de_1 + \sqrt{1-d^2}\,e_2\right) - \frac{ce_1^2+e_2^2}{2}.$$

If the signal has a positive impact on the agent's income ($w_1 > 0$), the maximisation problem of the agent is concave, and he chooses the following efforts in dependence of the wage rate w_1:

$$e_1^{\mathrm{SB}}(w_1) = w_1\frac{d}{c} \quad \text{and} \quad e_2^{\mathrm{SB}}(w_1) = w_1\sqrt{1-d^2}. \tag{2.21}$$

For a congruent signal ($d = \delta$) and full internalisation ($w_1 = 1$), the agent chooses the first-best efforts. But the principal has to keep the costs of uncertainty in mind when she selects the wage rate w_1 to maximise the second-best surplus:

$$\begin{aligned} B(e(w_1)) - C(e(w_1)) - C^{\mathrm{U}}(e(w_1)) &= \delta e_1(w_1) + \sqrt{1-\delta^2}\,e_2(w_1) \\ &\quad - \frac{ce_1^2(w_1)+e_2^2(w_1)}{2} - \frac{r}{2}w_1^2\sigma^2. \end{aligned}$$

By replacing $e(w_1)$ with $e^{\mathrm{SB}}(w_1)$, the objective function becomes:

$$\underbrace{\delta w_1 \frac{d}{c} + \sqrt{1-\delta^2} w_1 \sqrt{1-d^2}}_{\text{benefit}} - \underbrace{w_1^2 \frac{c\frac{d^2}{c^2} + (1-d^2)}{2}}_{\text{labour costs}} - \underbrace{w_1^2 \frac{r\sigma^2}{2}}_{\text{uncertainty compensation}}. \tag{2.22}$$

This function is concave and is maximised by setting the wage rate w_1 to

$$w_1^{\mathrm{SB}} = \frac{d\delta + \sqrt{1-d^2}\sqrt{1-\delta^2} c}{d^2 + (1-d^2)c + r\sigma^2 c}.$$

The second-best surplus is obtained by replacing w_1 in (2.22) by w_1^{SB}:

$$\Psi_d^{\mathrm{SB}} = \frac{\left[\frac{d}{c}\delta + \sqrt{1-d^2}\sqrt{1-\delta^2}\right]^2}{2(\frac{d^2}{c} + (1-d^2) + r\sigma^2)}.$$

The second-best surplus depends on d, which describes the composition of the signal. This composition is related to Baker's congruity measure: If $d = \delta$, the angle between the benefit coefficient vector $\boldsymbol{\delta} = (\delta, \sqrt{1-\delta^2})$ and the signal coefficient vector $\mathbf{d} = (d, \sqrt{1-d^2})$ is zero. Consequently, no misallocation should arise from using such a signal. If we hold the effect of uncertainty constant by fixing the signal-noise ratio, we expect congruent signals to deliver the best allocation and hence the highest surplus.

To verify this conjecture, we determine the signal composition d^*, which maximises the second-best surplus Ψ_d^{SB} while fixing the signal-noise ratio. A respective formula can be found in Schnedler (2003b) and leads to:

$$d^* = \delta \cdot \frac{(1 + r\sigma^2)}{\sqrt{\delta^2(1+r\sigma^2)^2 + (1-\delta^2)(1+cr\sigma^2)^2}}$$

The congruent signal δ only maximises surplus ($d^* = \delta$) if the second factor is one. The only admissible parameter constellation for which this is the case are $c = 1$ or $r\sigma^2 \to 0$. In other words, the congruent signal only maximises the second best surplus when either costs are identical for both efforts or when uncertainty does not matter. In more general settings, there are signals with the same signal-noise ratio which outperform congruent signals. Because the surplus-maximising signal has the same signal-noise ratio, it can only outperform the congruent signal with respect to effort misallocation. In other words, the allocation of effort induced by this signal creates a higher surplus than the allocation induced by the congruent signal. Consequently, congruent signals do not necessarily lead to the lowest misallocation and misallocation measures based on this assumption are problematic. More specifically, the congruity measure of Baker (2002) only identifies the least-biasing signal correctly when both efforts cost exactly the same ($c = 1$). Otherwise, the effort allocation of a congruent signal can be improved using a signal which emphasises efforts differently than the benefit function.

The example of congruent but biasing signals has an important practical implication. Suppose a firm is interested solely in the shareholder value. Then, paying workers in dependence of the shareholder value does generally not lead to the correct allocation of effort between tasks of these workers. The reason is that the preference of workers for particular tasks influences the allocation. The shareholder value reflects the interest of the firm but does not sufficiently counterbalance workers' preferences.

Discussion

How should we measure misallocation? Which of the three approaches should be used? What are the advantages of which approach? A measure of misallocation should relate characteristics of signals to their consequences on the allocation of effort between tasks. Possibly, the measure should work in a large class of incentive problems. It can be an ordinal measure, where only the ranking of signals matters, or preferably, a cardinal measure where differences in values can be interpreted in some meaningful way. In addition, it is helpful if the measure is easy to compute. In comparison to misallocation costs, which involve solving several maximisation problems, the measuret's of congruity by Feltham and Xie (1994) and Baker (2002) are relatively simple and thus preferable.

On the other hand, the class of incentive problems covered by these definitions is rather small. They are confined to linear signals and benefits, while the misallocation definition proposed here requires only that a unique solution to the incentive problem exists. For the sake of simplicity, most of the examples given in the following are using linear benefit, linear signals, and linear contracts; however, this does not mean that the definition is limited to these applications. Moreover, the values attached to signals by the two measures of congruity only inform about the relative misallocation, whereas the concept of misallocation costs is absolute and can be interpreted in a natural way as the willingness to pay for the respective signal.

At first glance, it seems to be an advantage that the measures of congruity rest only on the benefit and signal functions and are independent from the incentive scheme at work. In fact, it is this feature which makes them comparably easy to compute. However, for both measures this advantage turns into a shortcoming. According to Feltham and Xie's measure, congruity decreases when a signal is multiplied with a fixed number, although such a multiplication can fully be accommodated by simply dividing the signal through the fixed number before it is utilised in the incentive scheme. Thus, Feltham and Xie's measure detects differing congruity of two signals while they lead to perfectly identical effort allocation. As seen in the preceding section, Baker's measure is plagued by a similar problem. It produces a ranking of signals according to their congruity which does not coincide with the attractiveness of these signals: According to his definition, a biasing signal is more attractive for a mechanism designer than a balancing signal, everything else being equal. Hence, Feltham and Xie's as well as Baker's way of assessing misallocation is not (always) related to the surplus generated by the signal. Misallocation costs do not have this disadvantage: All else being equal, higher misallocation costs lead to a lower value of the signal.

2.3 Relationship between Misallocation and Uncertainty

This section examines the causes of misallocation and relates them to uncertainty. Is there a trade-off between uncertainty and effort allocation? Section 2.3.1 presents an example where the principal trades in accuracy for not getting exactly what she wants. Are uncertainty considerations always at the heart of misallocation? Section 2.3.2 considers a completely deterministic environment and demonstrates that misallocation emerges even when uncertainty is absent. Finally, Section 2.3.3 relates the trade-off between misallocation and uncertainty to the trade-off between bias and variance in statistics.

2.3.1 Uncertainty-induced Misallocation

Why is there a misallocation of effort when the latter is not contractible? One reason may be that the variance of signals associated with different dimensions of the effort vector are not the same. Then, asking for less effort along certain dimensions allows the principal to save on uncertainty costs.

As an example, consider again two efforts $e = (e_1, e_2)'$ and the same utility, costs, benefits, and outside option as in the example in Section 2.2.5 with the cost parameter $c = 1$. Differently from the example above, there are two independent input signals S_1 and S_2 with $S_i = e_i + \epsilon_i$ and $\epsilon_i \sim I\!N(0, \sigma_i^2)$ for $i = 1, 2$. Again, attention is limited to linear contracts, which now encompass a base wage w_0, a wage rate w_1 for the first signal realisation s_1, and a wage rate w_2 for the second signal s_2:

$$\Upsilon(s_1, s_2) = w_0 + w_1 s_1 + w_2 s_2.$$

Once more, the base wage w_0 is chosen so that the agent is just willing to participate, which implies that the principal has to pay uncertainty compensation, which increases with the wage rates of the signals: $C^{\mathrm{U}} = \frac{r}{2}(w_1^2\sigma_1^2 + w_2^2\sigma_2^2)$. These costs are deducted from the real surplus in the maximisation problem which the principal faces:

$$\max_{w=(w_1,w_2)'} B(e^{\mathrm{SB}}(w)) - C(e^{\mathrm{SB}}(w)) - \underbrace{\frac{r}{2}(w_1^2\sigma_1^2 + w_2^2\sigma_2^2)}_{C^{\mathrm{U}}\text{ uncertainty compensation}}$$

$$\text{such that} \quad e^{\mathrm{SB}} = (e_1^{\mathrm{SB}}, e_2^{\mathrm{SB}}) \in \operatorname{argmax}_{e_1,e_2} w_1 e_1 + w_2 e_2 - C(e(w)).$$

The solution to this problem is a wage scheme which induces second-best efforts equal to the respective wage:

$$e_1^{\mathrm{SB}}(w) = w_1^{\mathrm{SB}} = \frac{\delta}{1 + r\sigma_1^2} \quad \text{and} \quad e_2^{\mathrm{SB}}(w) = w_2^{\mathrm{SB}} = \frac{\sqrt{1-\delta^2}}{1 + r\sigma_2^2}.$$

The labour costs incurred by the worker for the second-best effort are:

$$C(e^{\mathrm{SB}}) = \frac{1}{2}\left(\left(\frac{\delta}{1 + r\sigma_1^2}\right)^2 + \left(\frac{\sqrt{1-\delta^2}}{1 + r\sigma_2^2}\right)^2\right).$$

Would this money be spent differently if there were no misallocation due to uncertainty?

To find the balanced effort allocation, the expenditure on real labour costs is fixed and can now be spent freely on the two efforts:

$$\max_{e=(e_1,e_2)'} B(e) - C(e) \text{ such that } C(e) = C(e^{\mathrm{SB}}). \tag{2.23}$$

If the constraint is written out, e_2 can be expressed in terms of e_1 and replaced in the objective function. The latter is then maximised in e_1.

This yields the balanced second-best effort as a function of effort costs $C(e^{\mathrm{SB}})$:

$$e_1^{\mathrm{BSB}} = \sqrt{\frac{\delta^2}{c}\frac{2C(e^{\mathrm{SB}})}{(1-\delta^2)c+\delta^2}} \text{ and } e_2^{\mathrm{BSB}} = \sqrt{(1-\delta^2)\,c\frac{2C(e^{\mathrm{SB}})}{(1-\delta^2)c+\delta^2}}, \tag{2.24}$$

where the second component uses the remainder of the budget and has been be retrieved from the side condition. Recalling that the costs were assumed identical here ($c = 1$) leads to:

$$e_1^{\mathrm{BSB}} = \delta\sqrt{2C(e^{\mathrm{SB}})} \text{ and } e_2^{\mathrm{BSB}} = \sqrt{1-\delta^2}\sqrt{2C(e^{\mathrm{SB}})}.$$

The balanced second-best efforts can be used to compute misallocation costs which are unique in this context because the second-best effort is unique:

$$\begin{aligned}
C^{\mathrm{UM}} &= B(e^{\mathrm{BSB}}) - B(e^{\mathrm{SB}}) \\
&= \delta(e_1^{\mathrm{BSB}} - e_1^{\mathrm{SB}}) + \sqrt{1-\delta^2}(e_2^{\mathrm{BSB}} - e_2^{\mathrm{SB}}) \\
&= \sqrt{2C(e^{\mathrm{SB}})} - \left(\frac{\delta^2}{1+r\sigma_1^2} + \frac{1-\delta^2}{1+r\sigma_2^2}\right) \\
&= \sqrt{\left(\frac{\delta}{1+r\sigma_1^2}\right)^2 + \left(\frac{\sqrt{1-\delta^2}}{1+r\sigma_2^2}\right)^2} - \left(\frac{\delta^2}{1+r\sigma_1^2} + \frac{1-\delta^2}{1+r\sigma_2^2}\right).
\end{aligned}$$

These costs become zero when there is no uncertainty about the signal ($\sigma_i = 0, i = 1, 2$) or when the Arrow-Pratt measure of risk-aversion tends towards zero ($\lim r \to 0$), hence they are indeed only caused by uncertainty. This justifies the use of the symbol C^{UM}, which designates uncertainty-induced misallocation costs.

Moreover, if the variances of both signals are equal, the costs of uncertainty imposed by using either of the two input signals are identical, and there is no reason to prefer one of the efforts based on the associated uncertainty. As a result, efforts are not distorted and misallocation costs are zero. Figure 2.3 depicts the situation of equal variances. To study the effect of rising uncertainty on effort allocation, the common variance is increased. Because the actual effort costs $C(\cdot)$ are symmetric, the overall costs are also symmetric, and the ratio of optimal inputs is identical for all levels of uncertainty. There is no misallocation of effort.

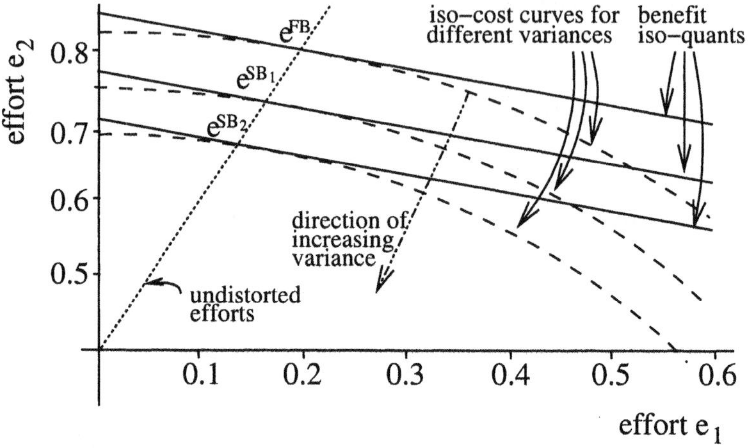

Fig. 2.3. Balanced allocation

Both efforts are measured with the same variance. If the variance is increased, the effort along both dimensions is reduced ($e_j^{FB} > e_j^{SB_1} > e_j^{SB_2}$, $j = 1, 2$)*, but the effort ratio remains constant.*

If, however, the signal of one effort, say e_2, has a larger variance, it is optimal to choose a higher relative level of effort e_1 in comparison to e_2. To visualise this effect, the variance in Figure 2.4 is chosen larger for the second effort signal. Consequently, the ratio between first and second effort is tilted in favour of the first effort.

The principal is willing to pay for a signal which leads to the same uncertainty compensation as in the second-best case, but allows for a free allocation of effort. Since the uncertainty compensation is the same, the costs of the new signal and the costs of the biasing incentive scheme are identical at the second-best effort allocation, and the respective iso-cost curves intersect. Using the new signal, the principal can reach a higher benefit, that is, the relevant benefit iso-quant now intersects with the vertical axis at about 0.2 instead of 0.12. The difference of about 0.08 is the willingness to pay for the balancing signal. The ratio of inputs when using the balanced signal is identical to the ratio of the first-best allocation or any other allocation where uncertainty does not affect allocation.

Note that the principal might have avoided the misallocation and induced the first-best input ratio even without access to a balancing signal. This could have been achieved by increasing the importance of

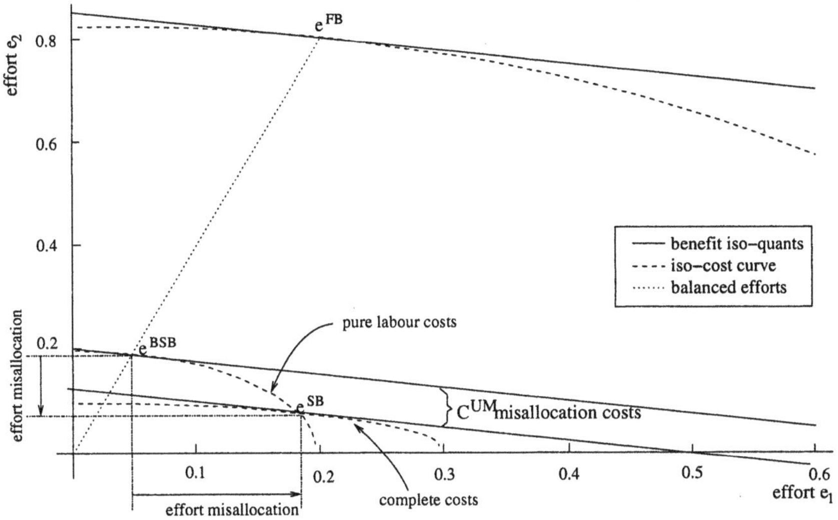

Fig. 2.4. Uncertainty-induced misallocation

In comparison with the first-best (FB), the second-best effort ratio is tilted in favour of the first effort (SB). The balanced effort allocation (BSB) requires more of the effort which is not so well observed (e_2) and less of the well observed effort (e_1).

the first signal, w_1, relative to the importance of the second signal w_2. Such an incentive scheme, however, would not have been second-best because a high w_1 requires uncertainty compensations that exceed the gains from a balanced input ratio; the misallocation of effort is voluntary and optimal because it allows to save on uncertainty compensation.

This example is a good illustration for the real life observation that supervisors prefer observable to unobservable effort, although the latter is more productive. The presence of workers at their workplace is often highly regarded even when they are idle and when they could be more productive at home. For example, in the first-year of the existence of the Bonn Graduate School (which the author had the pleasure to attend), the delivery of computers was delayed; most of the students had computers at home which they could use. Nevertheless they were required to "work" in their office. The less productive but less uncertain effort of sitting in an office apparently seemed more desirable than the more uncertain and more productive effort of letting the students work at home. The fact that it is the principal who is subject to risk-aversion in this example does not change the main conclusion: Uncertainty considerations lead to misallocated effort.

However, the assumptions and conclusions from the Bonn Graduate School example can be challenged. First, the requirement to be present might not have been second-best but simply irrational. Second, the sitting around might not have been so inefficient: For example, the resentment to sit around idly might provoke an interest in incentive schemes and lead to dissertations on this topic.

2.3.2 Misallocation Without Uncertainty

In the last subsection, the desire for low uncertainty costs led to a misallocation of effort; risk-aversion and uncertainty were necessary conditions for the misallocation to occur. Can misallocation also arise if signals are not noisy or if neither principal nor agent are risk-averse? Earlier, the theoretical concept of feasibility-induced misallocation costs was defined. Are these misallocation costs more than a technical definition, and do they emerge under some conditions? Both questions will be answered affirmatively in this section by means of an example.

Consider yet another variation of the previously introduced setting. Again, there are two efforts, benefit is linear in efforts, costs are quadratic, the outside option of the agent is zero, and the class of considered contracts is linear in the signal. The setting is almost identical to that in Section 2.2.5 where the existence of congruent but biasing signals was proven. There is only one deviation: The expected value of the signal $\tilde{S}$ is now contractible. Hence, the incentive scheme $\Upsilon(s) = w_0 + w_1 s$ can condition transfers on the values s of the deterministic signal $\mathrm{E}[\tilde{S}] = S(e) = de_1 + \sqrt{1-d^2}e_2$, and thus uncertainty is eliminated. Accordingly, we are dealing with the certainty version of the second-best programme from Section 2.2.5 and any misallocation must result from feasibility and not from uncertainty considerations.

Like in the previous examples, the base wage w_0 is set to the smallest level for which the agent is willing to participate and the base wage is ignored in the following. To determine the optimal performance wage rate w_1, the principal has to solve the following second-best maximisation programme:

$$\max_{w_1} B(e^{\mathrm{SB}}(w_1)) - w_1 S(e^{\mathrm{SB}}(w_1)) \quad \text{such that}$$
$$e^{\mathrm{SB}} = (e_1^{\mathrm{SB}}, e_2^{\mathrm{SB}}) \in \arg\max_{e_1,e_2} w_1\, d\, e_1(w_1) + w_1 \sqrt{1-d^2}\, e_2(w_1) - C(e) \tag{2.25}$$

where the constraint ensures that it is in the agent's interest to choose e^{SB}. Again, this effort choice will depend on the performance wage rate w_1. Since the maximisation problem of the agent is concave, the constraint can be replaced by the first-order conditions:

$$w_1 = \frac{e_1 c}{d} \quad \text{and} \quad w_1 \sqrt{1-d^2} - e_2 = \frac{e_1 c\sqrt{1-d^2}}{d} - e_2 = 0.$$

The strict functional relationship between the two efforts which is expressed by the right-hand side condition has its reason: The signal does not allow to disentangle the two efforts, and the incentive scheme cannot discriminate between them. Consequently, not all combinations of effort are feasible; the set of feasible efforts is limited:

$$E^{\mathrm{F}} = \left\{ \begin{pmatrix} e_1 \\ e_2 \end{pmatrix} \middle| \frac{e_1 c\sqrt{1-d^2}}{d} = e_2 \right\}.$$

Because the first effort component e_1 and the wage rate w_1 are uniquely related, the maximisation problem can be formulated as a choice over efforts:

$$\max_{e} B(e) - C(e) - C^{\mathrm{F}}(e),$$

where feasibility of effort is ensured by $C^{\mathrm{F}}(e)$, which is defined to be zero for all $e \in E^{\mathrm{F}}$ and prohibitively large otherwise. From this notation, it is apparent that we are dealing with the certainty programme, which results when surpressing the uncertainty in the incentive problem from Section 2.2.5 Solving the certainty programme leads to the effort allocations:

$$e_1^{\mathrm{C}} = \frac{d}{c} \frac{\left(\delta d + c\sqrt{1-\delta^2}\sqrt{1-d^2}\right)}{d^2 + c(1-d^2)} \quad \text{and}$$
$$e_2^{\mathrm{C}} = \sqrt{1-d^2} \frac{\left(\delta d + c\sqrt{1-\delta^2}\sqrt{1-d^2}\right)}{d^2 + c(1-d^2)}. \tag{2.26}$$

How much are these efforts different from the first-best efforts? The first-best efforts, which result from maximising $B(e) - C(e)$, are: $e_1^{\mathrm{FB}} = \frac{\delta}{c}$ and $e_2^{\mathrm{FB}} = \sqrt{1-\delta^2}$. They coincide with the certainty efforts, if the signal is congruent ($d = \delta$). In this case, the desired effort allocation can be achieved by the signal. Otherwise, the desired allocation is out of reach and the principal responses by reducing and reallocating effort. Because no uncertainty is involved the loss when using the signal instead of directly contracting on effort is entirely due to feasibility. These feasibility costs are computed as the difference between the first-best and the certainty surplus:

$$C^{\mathrm{F}} = \Phi^{\mathrm{FB}} - \Phi^{\mathrm{C}} = \frac{\delta^2 + c(1-\delta^2)}{c} - \frac{\left(\delta d + c\sqrt{1-\delta^2}\sqrt{1-d^2}\right)^2}{2c(d^2 + c(1-d^2))}$$
$$= \frac{1}{2}\frac{\left(\delta\sqrt{1-d^2} - d\sqrt{1-\delta^2}\right)^2}{d^2 + c(1-d^2)}.$$

The costs are zero if the signal is congruent. For non-congruent signals, efforts are reduced and misallocated and accordingly there are reduction and misallocation costs.

To determine misallocation costs, we compute the effort costs in the certainty programme:

$$C(e^{\mathrm{C}}) = \frac{1}{2c}\frac{\left(\delta d + c\sqrt{1-\delta^2}\sqrt{1-d^2}\right)^2}{d^2 + c(1-d^2)}.$$

Next, we allow the principal to spend these costs on effort without having to respect feasibility. This leads to the balanced certainty programme:

$$\max_{e=(e_1,e_2)'} B(e) - C(e) \text{ such that } C(e) = C(e^{\mathrm{C}}).$$

With the exception of the level of effort costs $C(e^{\mathrm{C}})$, this programme is completely identical to the second-best programme (2.23) from the previous section; the solution, the optimal choice of the first effort component, was described in (2.24). By replacing the effort costs in the formula with the effort costs from this section, the balanced certainty efforts can be computed:

$$e_1^{\mathrm{BC}} = \delta\frac{1}{c}\frac{\delta d + c\sqrt{1-\delta^2}\sqrt{1-d^2}}{\sqrt{(d^2 + (1-d^2)c)(\delta^2 + (1-\delta^2)c)}} \quad \text{and}$$
$$e_2^{\mathrm{BC}} = \sqrt{1-\delta^2}\frac{\delta d + c\sqrt{1-\delta^2}\sqrt{1-d^2}}{\sqrt{(d^2 + (1-d^2)c)(\delta^2 + (1-\delta^2)c)}}. \tag{2.27}$$

Comparing (2.27) with (2.26), it is clear that the balanced effort allocation is different from the second-best allocation, which demonstrates that misallocation can emerge in a certain environment. The respective misallocation costs are thus entirely induced by feasibility. They amount to:

$$\begin{aligned} C^{\mathrm{FM}} &= B(e^{\mathrm{BC}}) - B(e^{\mathrm{C}}) \\ &= \frac{\delta d + c\sqrt{1-\delta^2}\sqrt{1-d^2}}{c(d^2 + c(1-d^2))} \cdot \\ &\quad \left(\sqrt{\delta^2 + c(1-\delta^2)}\sqrt{d^2 + c(1-d^2)} - \delta d - c\sqrt{1-\delta^2}\sqrt{1-d^2}\right). \end{aligned}$$

If signal and benefit function coincide, balanced efforts and second-best efforts are identical and misallocation costs are zero.

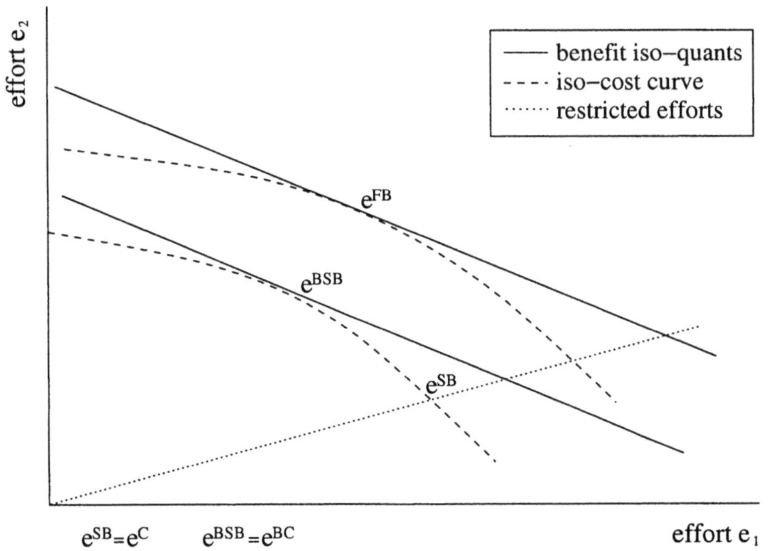

Fig. 2.5. Misallocation without uncertainty
The composition of the signal restricts the second-best allocation to the dotted line and separates it from the balanced allocation.

Figure 2.5 illustrates the first-best, second-best, and balanced allocation when there is no uncertainty. The signal composition restricts effort allocations to the dotted line. Hence, the second-best must be chosen from that line. Keeping costs constant by moving on the iso-cost curve, we find the balanced effort allocation where the benefit tangent touches the iso-cost curve. Revoking the restriction of having to spent a fixed amount on effort costs, the first-best benefit iso-quant can be reached.

In conclusion, effort misallocation and effort reduction are not necessarily uncertainty-induced; these effects also arise when signals do not allow to discriminate between different effort components.

2.3.3 Bias and Uncertainty in Statistics

Tradeoffs between misallocation and uncertainty occur if the use of a signal reduces misallocation but increases uncertainty. Then, the principal has to compromise between an incentive scheme which does not use the signal and imposes small uncertainty on the agent and an incentive scheme which utilises the signal and leads to low misallocation. Later, a fully fledged model with a misallocation-uncertainty trade-off will be developed. In this section, however, the relationship between such a trade-off and statistics is outlined.

To illustrate the trade-off between misallocation and uncertainty, imagine two incentive schemes: One scheme in which the principal gets what she wants, but at the price of a high uncertainty imposed on the agent, and a second scheme in which the principal does not exactly get what she wants, but which only imposes low uncertainty on the agent. Suppose that the features of the two incentive schemes stem from two monitoring techniques which render the agent's actions visible to a third party and thereby make them contractible. Then, the properties of the first scheme might result from a monitoring technique which gives on average the correct information about the agent's actions but this information is blurred by a noise term with an expected value of zero. On the other hand, the second incentive scheme could be based on a monitoring technique which does not measure the relevant actions of the agent but something related and these measurements are very precise.

In this example, the choice of an incentive scheme boils down to the decision between two types of monitoring, that is, information provision techniques. In statistics, estimators are used to provide information on unknown parameters and the two properties of monitoring techniques can be associated with characteristics of estimators: Measuring the action of the agent on average correctly can be associated with the unbiasedness of an estimator while precision is linked to its variance.

Thus, the counterpart to the first incentive scheme is an unbiased estimator with high variance, and the counterpart to the second one a biased estimator with low variance. In the hidden action context, the optimal incentive scheme is determined by solving a maximisation problem which blends uncertainty and misallocation considerations. But how does a statistician approach the trade-off between variance and bias? What estimator should be chosen?

Just as there are costs resulting from the choice of a bad incentive scheme, there is a loss related to selecting a poor estimator. Presumably, the goal of estimation is to learn more about an unknown parameter θ, in which case the estimator $\hat{\theta}$ should be close to it. How should the distance be evaluated? A popular choice which renders larger deviations comparatively more costly than smaller deviations is the *quadratic loss function*:

$$L(\theta, \hat{\theta}) = (\hat{\theta} - \theta)^2.$$

Because the estimator depends on a random sample, it is a random variable and the loss L is also a random variable. If we want to evaluate estimators, a random variable is of no help; we need a constant criterion such as the expected loss. Summarising these considerations, the criterion to evaluate estimators becomes the expected quadratic loss. This criterion is well known under the name *mean-squared error*:

$$\mathrm{MSE}(\hat{\theta}) := \mathrm{E}\left[\left(\hat{\theta} - \theta\right)^2\right].$$

While an optimal incentive scheme is chosen according to the lowest agency costs, estimators may be selected to minimise the mean-squared error. A lot of room in this chapter was devoted to the decomposition of non-contractibility costs. Is there a meaningful decomposition of the mean-squared error and can the components be associated with distortion and uncertainty?

To decompose the mean-squared error, subtract and add the expected value of the estimator $\mathrm{E}(\hat{\theta})$ inside the parentheses and multiply out:

$$\mathrm{E}\left[\left(\hat{\theta} - \mathrm{E}[\hat{\theta}] + \mathrm{E}[\hat{\theta}] - \theta\right)^2\right] = \mathrm{E}\left[\left(\hat{\theta} - \mathrm{E}[\hat{\theta}]\right)^2\right] + \left(\mathrm{E}[\hat{\theta}] - \theta\right)^2,$$

where the mixed term cancels due to $\mathrm{E}[\hat{\theta} - \mathrm{E}(\hat{\theta})] = 0$. Recalling the definitions of variance and bias (see, for example, Mood, Graybill and Boes 1974), this can be rewritten as:

$$\mathrm{E}\left[\left(\theta - \hat{\theta}\right)^2\right] = \mathrm{VAR}(\hat{\theta}) + \mathrm{BIAS}(\hat{\theta})^2.$$

Thus, the mean-squared error can be decomposed into cost components which are similar to the components of agency costs: The variance is related to uncertainty, and the bias reflects the aspect that the estimator is distorted. Hence, using the mean-squared error to select estimators in statistics implicitly involves a trade-off between uncertainty and misallocation very much like the trade-off in hidden action models.

2.4 Summary

The informational value of signals in hidden action models lies in their potential to reduce agency costs. These costs can be attributed to two sources: uncertainty, which together with risk-aversion leads to uncertainty costs, and the inability to recover different effort components from joint signals, which results in feasibility costs. Uncertainty costs and feasibility costs both include effort misallocation and effort reduction costs. In addition, uncertainty costs encompass uncertainty compensation.

Separating the costs of non-contractibility into different components allows us to answer the question how and why a particular signal reduces these costs. Thus, the source of the value of a signal is identified; for example, using the signal may diminish uncertainty compensation or decrease misallocation. The following chapters are chiefly concerned with misallocation and with signals which reduce misallocation at the price of higher uncertainty compensation.

3

Firm Performance Measures

This chapter deals with incentive schemes which are based on firm performance measures. It addresses the question whether it makes sense to pay workers according to firm's performance. The answer will rest on a trade-off between misallocation and uncertainty. It is argued that firm performance measures reduce misallocation and associated costs while increasing uncertainty and thus entailing higher uncertainty compensation. Additionally, it is considered that firm performance measures are unattractive because they provoke free-riding. However, within hidden-action models free-riding does generally not occur. The related literature is reviewed in Section 3.1. Because the benefit usually occurs at the firm level, firm performance measures are closely related to benefit signals. Section 3.2 presents a model describing the vice and virtue of benefit signals and contrasts it with an alternative that features a trade-off between misallocation and uncertainty compensation. Finally, Section 3.3 explains why firm performance measures resemble benefit signals in the sense that they tend to reduce misallocation and increase uncertainty of the worker's income. This leads to the hypothesis that firm performance measures are more intensely employed if misallocation entails high costs. Then, empirical ways to test this hypothesis are discussed and Section 3.4 summarises the main ideas of this chapter.

3.1 Multiple Agents and Free-Riding

In the previous chapter, the value of information was related to misallocation costs and uncertainty compensation. Accordingly, the degree to which firm performance measures are used can be discussed with respect to these two aspects. However, the theoretical framework was limited to a single worker, while firm performance measures generally reflect the effort of many workers. Interaction between the workers may be an additional source of misallocation besides feasibility and uncertainty. In particular, workers could lower their effort and free-ride on their colleagues. The following two sections review the literature on free-riding in the hidden-action context with multiple agents. In Section 3.1.1, it is pointed out that the existence of an employer eliminates the free-rider problem. Section 3.1.2 shows that free-rider effects are absent even if only workers know their abilities and can thus pretend to be less capable of working hard in order to get smaller work loads. Overall and contradicting a common belief, there is no theoretical model explaining the existence of free-riding in hidden-action models.

3.1.1 Principal as Budget Breaker

Free-riding within hidden action models seems to be a well-explored topic. As early as 1982, Holmström addresses the question of potential under-provision of effort in teams. To explain how the problem of under-provision – if it exists – is alleviated, the literature draws on peer pressure or social norms. (see e.g. Kandel and Lazear 1992 or Huck, Kübler and Weibull 2001). However, it is necessary to first establish that under-provision of effort is a problem in multiple-agent hidden action settings.

Why should under-provision of effort arise? Suppose there is a team of workers which exert effort to produce a joint benefit. While efforts cannot be observed or contracted upon, the benefit is perfectly observable. If the final benefit is always shared completely amongst workers, then workers internalise only partially the consequences of their effort on output: The individual worker weighs the marginal costs of increasing effort against his individual share of the marginal increase in benefit, which is lower than the joint marginal increase in benefit. Not appreciating the beneficial effects of his effort on others, the worker exerts too little effort; effort provision is below the Pareto-optimal level.

Holmström (1982) formalises this argument and emphasises the importance of the requirement that the benefit is always completely shared amongst the team members (*budget-balancing*). Slightly misrepresenting one of the main themes of the paper, Holmström is often quoted to have proven free-riding effects in moral hazard models. For example in his book, Salanié (1998) writes in the section on multiple agents in moral hazard models (Section 5.3.5, p. 127):

> If the Agent works in a team and only the team's global production can be measured, then the Agent's wage can only depend on global production unless the Principal gets reports on each member's effort. This clearly may induce Agents to free-ride on the effort of others, as proved by Holmström (1982).

But Holmström's formalisation of the free-riding argument provides only a reference point, which he calls *partnership*. It is important for the argument that all parties have to provide non-contractible effort. If a party exists whose effort is contractible or does not enter the joint benefit, the free-riding problem dissolves. In hidden action models, this is the case: The principal's actions are not entering the benefit function, she can act as a *budget breaker*, and previously impossible incentive schemes become feasible. In particular, the principal can ask for up-front payments from the agents which are paid back to them if the joint benefit amounts to the Pareto-optimal level; otherwise, they are kept by the principal. With such a scheme even the first-best result can be obtained. Other schemes with an element of collective punishment work similarly: The threat of being penalised when failing to achieve the joint goal prevents workers from free-riding.

Schemes based on collective punishment cannot be employed when all agents influence the joint benefit in a non-contractible way because then it is not credible that the collective punishment is carried out. Finding themselves in a state of the world where they are supposed to penalise each other, they are likely to suspend the punishment because it harms everybody. Foreseeing that the punishment will be suspended, they again provide too little effort. When there is an external party which profits from the penalty, this party renders the threat credible because it will object the suspension of the punishment. If this party is not truly external and also has to provide a non-contractible Pareto-optimal effort contribution, she will not meet this obligation and chose a different effort so that the joint goal is not met and she can profit from the penalty.

Schemes in which all workers are punished are sometimes discarded with the argument that workers are financially constrained and by alluding to the harsh nature of collective punishment – Lazear calls them *Stalin schemes* (1995, p. 50). However, this argument only has bite if workers do not receive any base payment or if this base payment is low. Otherwise, the size of punishment could be subtracted from the base payment and be paid out in case the joint goal is met. Punishing then becomes rewarding and the workers previously terrorised by the menace of a collective punishment happily strive for a common cause. The next section (3.1.2) presents a context in which informational rents lead to a sufficiently high base payment so that a rewarding instead of a punishing scheme can be constructed even when workers have limited endowments or access to credit.

Holmström (1982) recognises that the presence of the principal eliminates the free-riding problem and presents respective mechanisms, but his analysis is placed in a setting in which the joint benefit is perfectly observable. To accommodate uncertainty considerations, Holmström then introduces noise and risk-aversion. The main conclusion remains robust: There are incentive schemes which can at least proxy the first-best result arbitrarily closely. This implies that free-riding is not present. Only if the agent is risk-averse with a utility which is unbounded from below when income decreases, the first-best cannot be achieved anymore. However, the failure to generate the first-best does not imply the presence of free-riding. A reduction in effort can already be observed in the hidden action model with a *single* agent. The question whether there is an additional reduction of effort resulting from free-riding in this particular case has not yet been explored.

Once signals are noisy and workers risk-averse, the value of additional signals can be discussed. Holmström (1982) devotes the remainder of his paper to the extension of the well-known and already mentioned information principle to multiple agents.

Generally, Holmström's aim is not to prove that free-riding prevails in a principal agent context but to show that the existence of a principal can help to solve the free-riding problem. Accordingly, Holmström considers the existence of a principal or owner to be a major advantage of capitalistic production in comparison with non-profit or labour managed firms (1982, p. 325):

> In fact that capitalistic firms feature separation of ownership and labor implies that the free-rider problem is less pronounced in such firms than in closed organizations like partnerships.

Note that Holmström is very careful in his wording and does not claim that the separation of ownership and labour, i.e. the introduction of a principal, solves the free-riding problem. Nevertheless, he shows that a principal can eliminate free-riding – even under risk-aversion and uncertainty.

3.1.2 When Costs Are Private Information

The bonus contract, which leads to the first-best solution under certainty and approximates the first-best solution under uncertainty, is based on punishing the agents when the joint outcome target is not reached. The optimal punishment might cut into the initial endowment of agents. In real-life, it is rare that incentive contracts stipulate fines which are to be paid by workers when output targets are not met. In addition, it is unlikely that real employers have the necessary information in order to construct the optimal incentive schemes. In particular, it is questionable whether they have the detailed knowledge about the costs which effort choices inflict on the workers. However, these costs are essential for the optimal incentive scheme. It is more realistic to assume that only the worker knows the effort costs. Then, the worker may pretend to suffer from effort exertion more than he actually does in order to reduce the contribution asked from him. Hence, the principal's ignorance might translate into free-riding. On the other hand, she may try to uncover the effort costs by screening workers.

McAfee and McMillan (1991) analyse whether the free-riding effect occurs when only workers know the costs of effort exertion. In their model, each worker is endowed with a cost parameter which is private knowledge and the distribution of which is commonly known. Two types of noise-free signals are considered: a collective benefit signal and signals for individual contributions. If a free-riding effect is present, the optimal incentive scheme using individual signals should outperform the optimal scheme employing the collective signal. Whatever signal the incentive scheme is based on, it confounds the ability of workers with the effort exerted by them. As a consequence, workers receive a rent, work less than when their costs were known, and the first-best effort cannot be achieved.

However, the central point of McAfee and McMillan (1991) is not that it is impossible to obtain the first-best effort when there is private information of workers; this result is well-known from adverse selection models. Rather, they inquire whether the confounding of abilities and efforts of *different* workers in a collective signal leads to an additional reduction in efficiency. Using the *revelation principle*, McAfee and McMillan (1991) search for the optimal incentive scheme based on the collective signal under the constraint that all workers have an incentive to report their cost parameter truthfully. The optimal contract consists of two parts: one part which is fixed and another part which rises linearly in the collective benefit signal. Before the exertion of effort, each worker announces his productivity truthfully, and these reports are used to calculate the expected benefit. Together with the actual benefit the calculated expected benefit determines the size of the variable pay: If the expected benefit is achieved, there is no additional payment; higher benefits are honoured, while lower benefits are punished. Differing from the optimal incentive schemes in the hidden action model devised by Holmström (1982), workers may have limited endowments or may be credit-constrained in McAfee's and McMillan's model; the informational rent which results from their superior knowledge about effort costs provides them with financial means. These means can be used to pay the principal if the actual benefit is below the expected benefit.

The efficiency loss resulting from the optimal incentive scheme based on the joint benefit amounts exactly to the rents which need to be paid in order to reveal the workers' cost parameters. Those costs, however, are also incurred when the incentive scheme is based on individual contributions. Hence, using individual information has no advantage over using collective information, and there is no free-rider effect resulting from the usage of the joint benefit measure. Adding signals about individual contributions to an optimal incentive scheme based on the joint benefit has no value.

Vanderveen (1995) introduces noise and risk aversion to the model of McAfee and McMillan (1991). Since additional signals reduce the uncertainty which is shouldered by the risk-averse worker, it is not surprising that adding individual signals leads to Pareto improvements. In the pareto-improving incentive scheme of Vanderveen (1995), the individual contributions remain unchanged. So, the improvement that he finds is entirely due to a reduction in uncertainty and not caused by less free-riding.

Consequently, the analysis of the model where costs are private information adds support to the conjecture that free-riding does not occur once there is a principal. The conjecture contravenes the common belief that free-riding is a serious issue in teams. To console this belief with the theoretical findings, one can assume that for some reason the respective optimal contract cannot be used in reality or that the presented models lack a detail which is relevant. Of course, it may also be that the common belief is wrong.

Summarising this section, there is no hidden-action model in which the reduction in effort can be attributed to free-riding. Hence, free-riding cannot be regarded as a disadvantage of using aggregate signals like firm performance measures in incentive schemes.

3.2 The Use of Benefit Signals

As will be argued later, firm performance measures and benefit signals share the property of reducing misallocation. Thus, the utilisation of benefit signals teaches a valuable lesson on the use of firm performance measures.

Recall that a benefit signal is a (possibly noisy) function of the benefit that the employer draws from the effort exerted by the worker. Hence, it reflects all dimensions of effort. Moreover, to produce the signal, efforts interact in the same manner as in the benefit function. This distinguishes benefit signals from other performance signals which inform only about some efforts or in which the relationship between efforts is different from the benefit.

Lazear (1995) devises a model to explain the usage of benefit signals, which will be explained in Section 3.2.1. In his model other performance signals appear only marginally, and the decision to use benefit signals is driven by an optimal allocation of workers to jobs; effort choice is not modelled. Section 3.2.2 offers an alternative: a hidden action model with an effort signal in addition to the benefit signal. In this model, the utilisation of the benefit signal is determined by a trade-off between misallocation and uncertainty. The model predicts that the intensity of benefit signals is dependent on the shape of the benefit function. By associating benefit signals with firm performance measures, this prediction is later adopted to the present context, operationalised and empirically tested.

3.2.1 Lazear's Input versus Output Measure Model

Performance measures can be classified in several ways. A popular classification is the distinction between input and output measures – see for example Lazear (1995) or Prendergast (2000). Unfortunately, no formal definitions exist for the two categories. Informally, input measures seem to be associated for example with working hours and the material which is consumed in the course of effort exertion, whereas output measures relate to some result of effort exertion. Given the loose definitions, it is difficult to determine whether input or output measures are more preferable and how this assessment changes under different conditions.

Nevertheless, Lazear (1995) devotes a section in the second chapter of his monograph on personnel economics to the topic "Payment by Input versus Payment by Output". The model which he devises in that section is relevant for the current analysis because it features a benefit signal. This signal informs employer and worker about the worker's productivity and Lazear defines this productivity to be identical to the benefit which the employer draws from the worker.

Before observing the signal, both employer and worker do not know the benefit. The benefit signal costs κ, and the question is whether it is better to buy the signal and learn the benefit or stay ignorant. A zero profit condition on the side of the employer guarantees that the worker is paid the expected benefit if the actual benefit is unknown and that the worker receives the benefit minus the costs κ if the benefit signal is bought. So far, the worker prefers ignorance: His expected payoff then amounts to the expected benefit; otherwise it is at least κ lower. When the worker is risk-neutral, as supposed by Lazear, the payoff under ignorance exceeds that of knowing the benefit exactly by κ.

To render the knowledge about the actual benefit more valuable, Lazear assumes that there exists an alternative job, which yields the fixed benefit ω to the employer. Due to the zero-profit condition, the wage for the worker at this job is ω. Hence, it is optimal for the worker to switch to the alternative job when his benefit at the initial job is below ω. The existence of the alternative job complicates the worker's decision whether to buy the benefit signal.

This decision can be framed in terms of a lottery: The worker has to decide whether to buy a lottery ticket which costs κ, that is whether to purchase the benefit signal. All tickets in this lottery win and the minimum prize is ω because the worst which can happen to the worker is that he gets the alternative job. The guarantee to get at least ω censors the prize distribution from below, so that the lottery may be called a *censored lottery*. Not surprisingly, a higher guaranteed minimum winning prize ω makes the lottery more attractive to the worker, while higher ticket costs κ render it less attractive (see Figure 3.1). These are the two essential findings of Lazear.

Lazear additionally parameterises the costs in the following way: $\kappa := \frac{\tau}{1-\lambda}$, where Lazear's τ stands for the nominal costs of the signal (and is completely unrelated to the meaning of τ in this text) and λ describes the share of the work life spent in form of a probation period; this probation period elapses before the benefit signal is observed. Thus, the actual costs of the benefit signal increase in nominal costs and length of the probation period. Consequently, the value of the benefit signal is diminished either if its nominal costs increase or if the employer has to wait longer before she can use the signal.

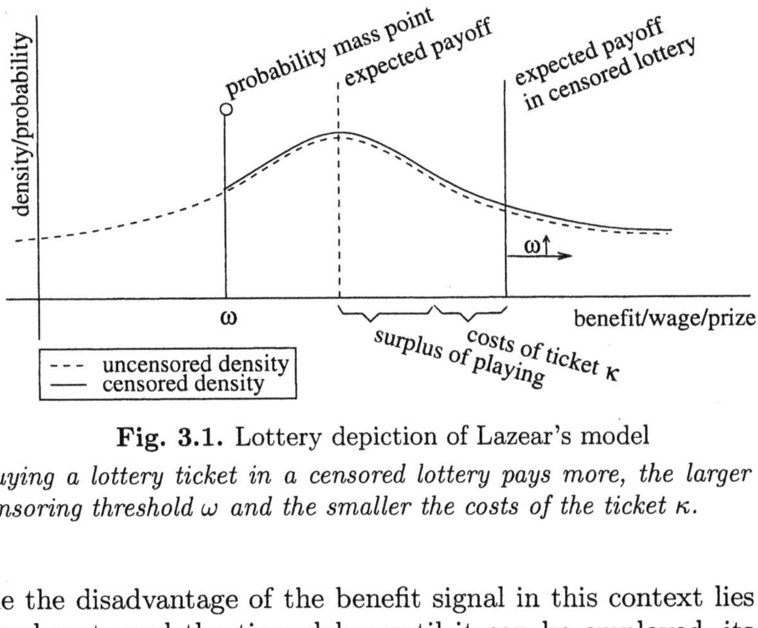

Fig. 3.1. Lottery depiction of Lazear's model

Buying a lottery ticket in a censored lottery pays more, the larger the censoring threshold ω and the smaller the costs of the ticket κ.

While the disadvantage of the benefit signal in this context lies in its nominal costs and the time delay until it can be employed, its value is related to the existence of the alternative job with the fixed benefit ω. Knowing the information conferred by the benefit signal enables the employer to better allocate workers to jobs. Lazear (1995, p. 20) expresses this idea in the following way:

> The advantage of undertaking measurement and paying the corresponding piece rate is that low-quality workers are sorted out to a higher-valued use.

Lazear calls the benefit signal "output measure", and although he strives to explain when payments are based on output rather than input measures, the latter are not explicitly modeled. Thus, the model explains advantages and disadvantages of the output measure, but remains mute on the properties of input measures. The only verbal indication about the comparative advantages of input measures which Lazear provides is the following (1995, p. 20):

> There are a number of factors that affect a firm's choice between payment on input versus payment on output. One important factor is measurement cost. It is often easier to obtain a rough measure of input than it is to obtain an balancing measure of output.

The term "measurement cost" is relatively vague, but the mention of "rough measures of input" in contrast to "unbiased measures of output" suggests that misallocation plays a role for the decision between input and output measures. Thus, the decomposition of costs into misallocation costs and other cost components might provide fruitful to explain this decision. Lazear does not explore this theme further, and it does not enter his model. But it will be taken up in the next section (3.2.2).

The effect of sorting, which generates the value of benefit signals, has already been analysed by Lazear in an earlier article (1986). The essence of both models is that workers self-select into jobs with differing incentive schemes according to individual characteristics. Like effort exertion, sorting is no end in itself, but only a means to increase the surplus of the interaction between employer and worker. It is an open question whether performance measures are in practice predominantly used to improve allocation of workers to jobs or to induce effort. With respect to firm performance measures, however, there is less ambiguity. Within one firm, firm performance is by definition the same for all workers. Thus, firm performance measures cannot be used to directly allocate workers to jobs. On the other hand, the employer could vary the job allocation systematically, measure firm performance, and thereby indirectly determine which allocation is best. While this is theoretically possible, it seems to have little practical relevance: The quality of the allocation of workers to jobs is probably better assessed by a subjective report of a supervisor than by a firm performance measure.

To some degree, the question whether performance measures are employed to sort workers to jobs or to induce workers to exert effort is sophistic. Ultimately, some sort of effort has to be exerted by the worker to generate benefit. Thus, the effect of benefit signals on effort is worthwile to examine. But effort choices are completely absent from the model of Lazear which was discussed here. The following section presents a complementary model which encompasses the effort choice of the worker and in which payments can be conditioned on a benefit and an effort signal. If the effort signal is identified as an input measure and the benefit signal as an output measure, then this model describes the choice between "Payment by Input versus Payment by Output", which is mentioned in the section title of Lazear's monograph.

3.2.2 A Misallocation-uncertainty Trade-off Model

The model of Lazear emphasises the use of benefit signals to sort workers. This section suggests that the function of benefit signals might also be the correct allocation of effort. According to this idea, benefit signals are used to prevent misallocation and avoid the respective costs. They are not excessively used because they tend to be rather noisy. Overall, there is a trade-off between misallocation and uncertainty which gives the following model its name: *misallocation-uncertainty trade-off model*.

This model is a simple hidden action model and draws from several examples encountered before. In Section 2.3.2, the employer was able to control the overall effort level of the worker by making the latter's wage dependent on a signal $S(e) = e_1 d + e_2\sqrt{1-d^2}$. Now, suppose that the signal composition takes the value $d = \sqrt{0.5}$. Then, the employer can multiply this signal by $\sqrt{2}$ in order to create a signal which measures the total input of the agent in terms of effort:

$$S_1(e) = e_1 + e_2.$$

This signal is free from noise so that using it entails no uncertainty costs. A possible interpretation of the signal S_1 arises when e_1 and e_2 represent the time which the worker spends two tasks. Then, the signal S_1 is simply the total working time. In addition, there is a noisy benefit signal available:

$$S_2(e) = B(e) + \epsilon \text{ with } \epsilon \sim I\!N(0, \sigma^2).$$

This signal could be interpreted as the shareholder value or any other firm performance signal which captures the interest of the employer but is contaminated with randomness beyond the worker's influence.

Like in all previous examples, the benefit should be a linear combination of the two efforts:

$$B(e) = \delta e_1 + \sqrt{1 - \delta^2} e_2 + \epsilon.$$

The costs of effort exertion are again quadratic. To avoid uncertainty-induced misallocation, we set costs to be equal for both efforts ($c = 1$), so that

$$C(e) = \frac{e_1^2 + e_2^2}{2}.$$

The critical character of this assumption was discussed, earlier: Under this assumption, congruent signals are balancing and uncertainty does not induce misallocation. By assuming equal costs, it is possible to isolate misallocation costs from uncertainty costs and clarify the trade-off.

The agent is risk-averse with the same utility function as before:

$$u(\iota) = -e^{-r\iota},$$

where r is the Arrow-Pratt measure of risk aversion and $\iota = \tau - C(e)$ is the income of the agent composed of received transfers minus effort costs.

Once more, only contracts are considered which linearly relate transfers to signal realisations (s_1, s_2):

$$\Upsilon(s_1, s_2) = w_0 + w_1 s_1 + w_2 s_2.$$

Via the wage rate w_1, the employer perfectly controls the total effort. The wage rate w_2 affects the effort ratio (in the right direction). Both coefficients w_1 and w_2 represent the sensitivity of transfers with respect to the appropriate signal. The linear contract space is particularly attractive in this case, because it allows us to assess the relative sensitivity of the benefit signal in comparison with the effort signal by the ratio $\frac{w_2}{w_1}$. Using the relative sensitivity, it becomes possible to analyse when benefit signals are relatively more important than effort signals.

The main disadvantage of the effort signal is misallocation. Unless both efforts have the same marginal effect on the benefit ($\delta = \sqrt{0.5}$), efforts are not balanced. The disadvantage of the benefit signal are the uncertainty costs. Figure 3.2 depicts the uncertainty costs of the benefit and misallocation costs of the effort signal. They give a rough idea which signal is cheaper for which parameter values: The costs of the benefit signal do not change in the marginal productivity of the first effort, while the costs of the effort signal are lowest for $\sqrt{0.5}$ and increase with the distance to that point. However, a change in σ increases the costs of the benefit signal and leaves those of the effort signal constant.

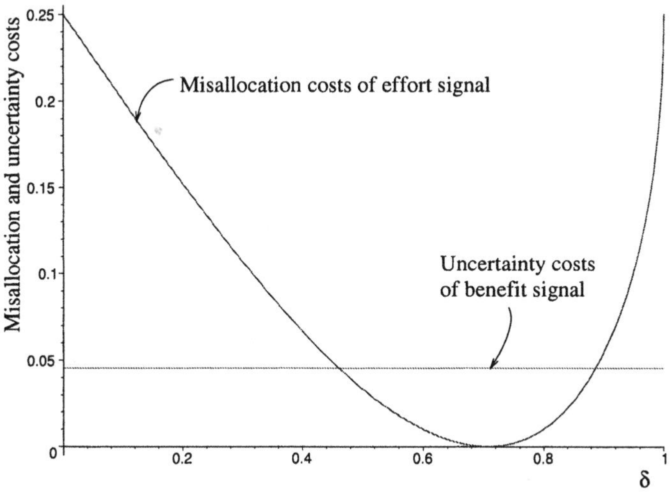

Fig. 3.2. Costs of benefit and effort signal
While the costs of the benefit signal are constant, the costs of the effort signal vary in the benefit parameter. The effort signal is congruent for $\delta = \sqrt{.5}$.

Even if benefit signals are in reality also leading to misallocation, it seems fair to say that their strength in comparison to effort signals is allocation and their weakness uncertainty. Effort signals such as hours of work often exhibit little unsystematic noise (low variance) but fail to measure the true objective (high misallocation), while benefit signals, e.g. profits, represent the objective well (low misallocation) but suffer from high uncertainty induced by factors beyond the worker's control like changing demand for the product (high variance). Again, it should be mentioned that if the effort signal is identified as an input measure and the benefit as an output measure, this model can be viewed as an alternative to the model of Lazear (1995, p. 20), which was presented in the previous section and which describes the choice between the input and the output measure. In contrast to Lazear, who only models the output measure, there are two signals here, and it can thus be analysed how the choice for a signal depends on its properties.

The maximisation programme in which the utilisation of the effort and the benefit signal is jointly determined has the form:

$$\max_{w_1,w_2} B(e_1^{\mathrm{SB}}, e_2^{\mathrm{SB}} - C(e_1^{\mathrm{SB}}, e_2^{\mathrm{SB}}) - \frac{r\sigma^2 w_2^2}{2} \quad \text{such that}$$

$$(e_1^{\mathrm{SB}}, e_2^{\mathrm{SB}}) \in \arg\max_{e_1,e_2} \quad w_1 e_1 + w_1 e_2 + w_2 \delta e_1 + w_2 \sqrt{1-\delta^2} e_2 - C(e). \tag{3.1}$$

To derive this programme, the base wage w_0 was chosen so that the participation constraint is binding. The side condition of the programme can be used to calculate the effort choice of the agent e^{SB} and to see how it depends on the incentive scheme (w_1, w_2):

$$e_1^{\mathrm{SB}} = w_1 + \delta w_2 \quad \text{and} \quad e_2^{\mathrm{SB}} = w_1 + \sqrt{1-\delta^2} w_2.$$

Thus, the wage rate of the effort signal w_1 equally increases both efforts, while the wage rate of the benefit signal w_2 leads to higher exertion of the more productive effort.

Replacing the calculated second-best effort e^{SB} in the objective function in (3.1) leads to a maximisation problem which only depends on the incentive scheme parameters. The resulting optimal choice of the signal coefficients is:

$$w_1^{\mathrm{SB}} = \frac{r\sigma^2(\delta + \sqrt{1-\delta^2})}{1 - 2\delta\sqrt{1-\delta^2} + r\sigma^2} \quad \text{and} \quad w_2^{\mathrm{SB}} = \frac{1 - 2\delta\sqrt{1-\delta^2}}{1 - 2\delta\sqrt{1-\delta^2} + r\sigma^2}.$$

The relative importance of the effort signal in comparison with the benefit signal is:

$$\frac{w_1^{\mathrm{SB}}}{w_2^{\mathrm{SB}}} = r\sigma^2 \frac{\delta + \sqrt{1-\delta^2}}{1 - 2\delta\sqrt{1-\delta^2}}.$$

To illustrate the trade-off between benefit and effort signal, the consequences of parameter changes on this quantity are examined. Recall that the effort signal leads to an equal split of effort among the tasks. The larger the difference in importance of the tasks, the further away is δ from $\sqrt{0.5}$, the equal split becomes less apt, and the misallocation costs of the signal increase. The principal reacts and reduces the misallocation costs by increasing the relative importance of the less biasing benefit signal. The price, a higher uncertainty in worker's wage, is more than offset by the gain, a higher benefit. Conversely, the closer

δ to $\sqrt{0.5}$, the less distortive is the effort signal and the larger will be its role in the wage bill. In the limit ($\delta \to \sqrt{0.5}$), the effort signal takes over and no benefit signal is used anymore.

If the uncertainty of the benefit signal σ^2 is increased, it is optimal to insure the agent better against uncertainty and to reduce the influence of the benefit signal on the wage. However, it is not a mere reduction which can be observed but rather a substitution of the benefit by the effort signal. The gains of higher effort exertion and a lower uncertainty premium outweigh the increase in misallocation costs resulting from this step. Conversely, a lower variance leads to more influence of the benefit signal in the incentive scheme. If the uncertainty considerations do not matter ($r\sigma^2 \to 0$), the benefit signal will be the only signal used.

In the model, the shape of the benefit function as captured by δ influences the value of benefit signals. As an example, consider a professor who is hired to teach and to do research. If teaching and research are perfectly substitutable in the sense that one hour of teaching creates the same benefit as one hour of research, then doing only research leads to no misallocation costs. However, if the benefit can only be obtained by a particular combination of research and teaching, then focusing on research has disastrous effects and misallocation costs are high.

The preceding model and example should convince the reader that the shape of the benefit function determines whether misallocations translate into misallocation costs. Consequently, the value of a signal which reduces misallocations depends on the shape of the benefit function. This finding is not only true for the particular example of a benefit signal but it holds more generally for all signals which help to diminish misallocation. It is argued in the next section that firm performance measures tend to reduce misallocation. Consequently, the value and utilisation of firm performance measures varies with the benefit function or – if the benefit results from production – with the production technology faced by the worker.

3.3 Implications and Implementation

The previous section dealt with a signal that reduced the misallocation of effort but increased the uncertainty imposed on the risk-averse worker. It has been shown that the intensity with which such a signal is employed varies with the need to decrease misallocation costs.

In this section, it is argued that firm performance measures also reduce misallocation and increase uncertainty. Hence, the result from the previous section can be applied to explain the intensity to which firm performance measures are employed. It is explained why the intensity can be proxied by firm performance pay and why the need to reduce misallocation is reflected by task characteristics. So that by the end of this chapter, the relation between the intensity of firm performance pay and the need to reduce misallocation can be empirically tested using firm performance pay and task descriptors.

Section 3.3.1 argues that firm performance measures tend to reduce the misallocation of an incentive scheme and to increase uncertainty about pay. Section 3.3.2 explains why firm performance pay proxies the intensity of firm performance measures. Section 3.3.3 suggests to measure misallocation costs by descriptions of the tasks which are to be accomplished by the worker.

3.3.1 Benefits and Costs of Firm Performance Measures

This section identifies the benefits and costs of firm performance measures and concludes with a hypothesis about the degree to which firm performance measures are employed in incentive schemes.

Previously, free-riding has been ruled out as a convincing disadvantage of firm performance measures because the theoretical evidence is not compelling. Furthermore, an improved allocation of workers to jobs was discarded as an advantage of firm performance measures in spite of its validity for other performance measures. What, then, are the benefits and costs of using firm performance measures?

First, it should be noted that the class of firm performance measures includes very different entities such as shareholder value, firm profits, firm sales, firm output, and other balance sheet information aggregated on the firm level. Thus, one cannot expect to find a precise list of benefits and costs which always fits all these variations. Even more frustrating, it has been shown earlier that the value of a signal depends also on the presence of other signals; the class of alternatives to firm performance measures is abundant. Considering the variability in the value of firm performance measures which results (i) from the variation in the availability of alternative performance measures and (ii) from differences between the firm performance measures, it seems almost impossible to derive robust conclusions. Nevertheless, the fact that all firm performance measures operate on the same aggregation level allows us to state properties which predominantly hold and which are sufficient to generate meaningful implications in the light of the misallocation-uncertainty trade-off model. Namely, the following paragraphs argue that firm performance measures reduce misallocation and increase uncertainty. Due to the general nature of the problem, the argument will remain verbal.

Misallocation Reduction by Firm Performance Measures

The objective of the employer is likely to be related to the firm as a whole. It is generally simpler to find balanced performance measures on the same aggregation level as the objective. If the performance measure of another less aggregated level is used, it is necessary to find out the implications of the global objective on this local level and to define the corresponding local objective, before a respective performance measure can be constructed. For example, increasing profit may be a goal of the employer which can be measured on the firm level. If the employer wants to use individual performance measures, she first has to determine how "increasing profits" translates into an objective for the worker. She must figure out what behaviour she desires from the worker before she is able to construct an incentive scheme based on individual performance measures. Otherwise, the incentive scheme is bound to be inapt; not necessarily because of Kerr's (1975) *folly of rewarding A while hoping for B* but because the employer does not even know what B she is hoping for. Simply because the employer knows more about the firm level than about the individual level, she is more capable of finding and constructing firm performance measures which represent her interests and which are consequently less biasing than other performance measures.

Firm performance measures have another feature which suggest that they are useful to reduce misallocation. They are sensitive to actions which are difficult to specify (even if they were known) or which cannot be foreseen when the working contract is written. For example, when a worker is approached by a coworker who has a problem, it may be advisable to help this coworker to achieve the overall objective, but it is difficult to foresee whether a problem will occur, who will be in need of help, and who should provide the help and to what extent. To specify all these contingencies and actions in advance may not be feasible. However, if the help boosts firm performance and workers are paid according to a firm performance measure, they will determine by themselves whether and whom to help. It is reasonable to suppose that this decision is closer to the interest of the employer than when there is no dependency of remuneration on firm performance.

To summarise, firm performance measures cover more dimensions of the worker's behaviour and they weigh these dimensions more appropriately than other performance measures. In other words, making the remuneration of a worker more dependent on firm performance reduces the misallocation in effort allocation. However, firm performance measures are not free from misallocation. Even in the extreme case that the objective coincides with a firm performance measure – for example, when the employer is only interested in maximising the shareholder value – it is an illusion to believe that firm performance measure will lead to a balanced effort allocation. As seen in Section 2.2.5, the coincidence of objective and signal leads to misallocation if uncertainty is involved. Suppose that firm performance measures induce a smaller misallocation than alternative performance measures. This suffices for firm performance measures to be successfully employed to reduce misallocation.

Increase of Uncertainty Due to Firm Performance Measures

If firm performance measures have so favourable qualities, why are they not always used to the largest possible extent? Section 3.1 questioned that free-riding is a serious obstacle to employ firm performance measures but Section 3.2 suggested another possible reason: high variability.

Firm performance measures depend on a relatively large number of determinants in comparison with less aggregated measures. The shareholder value, for example, is determined by the behaviour of other workers, traders at the stock market, management decisions, product market conditions, etc. All these factors are beyond the control of the worker. Moreover, they are unpredictable to some degree. Thus, there is variation which is unsystematic from the worker's perspective. Such variation might also plague more individualistic performance measures such as working hours. For example, the worker may oversleep the beginning of work and lose the wage of one hour. But the worker exercises considerably more control over most individual performance measures; in case of having overslept, he might even recoup the lost hour by working longer. Thus, the influence of the unsystematic component is much higher on firm performance than on individual performance. Respectively, firm performance measures exhibit a higher variance. Given risk aversion of the worker, this directly translates into higher uncertainty costs.

Utilisation of Firm Performance Measures

If firm performance measures reduce misallocation and increase the uncertainty imposed on the worker, they essentially relate to alternative performance measures in the same way like the benefit signal relates to the effort signal in the misallocation-uncertainty trade-off model. Hence, some findings of this model carry over. Similar to the benefit signal, firm performance measures are the more useful, the higher misallocation costs are. The extent to which they are employed should be positively related to the misallocation costs which were incurred in absence of the firm performance measures. Misallocation costs, in turn, depend on how different dimensions of effort relate to each other in the benefit function. If efforts are substitutable, misallocation does not lead to costs and firm performance measures should play a minor role in the incentive scheme.

These considerations, which result from the combination of the theoretical findings in the misallocation-uncertainty trade-off model with the assumption that firm performance measures decrease misallocation and increase uncertainty, can be summarised in the following way:

Hypothesis
The intensity with which firm performance measures are employed increases with need to reduce misallocation.

To identify the need to reduce misallocation, the substitutability of effort in the benefit function might provide an indicator.

3.3.2 Gauging the Intensity of Performance Measures

The intensity to which firm performance measures are employed within an incentive scheme is still an undefined concept. Before the hypothesis can be tested empirically, this concept must be defined and translated into a measurable quantity.

Intensity of Performance Measures

The influence of the performance measure on the behaviour of the worker may vary. For example in the misallocation-uncertainty trade-off model, the influence of the benefit signal on effort has been decreased by the employer when the variance of this signal increased. The benefit signal has been less intensively used, and this manifested itself in a smaller wage rate of this signal.

Does this mean that the wage rates of signals can be used to assess the intensity? The worker's behaviour is indeed more influenced by a signal when the respective wage rate is increased. But this idea to define intensity has two shortcomings. First, it can only applied to linear remuneration schemes where wage rates exist. Second, it is not robust to re-scaling: If the signal is divided by a number, the wage rate can be multiplied by the same number and changes, while the importance of the signal for the behaviour of the worker remains the same.

A more appropriate definition for the intensity with which a performance measure is used in an incentive scheme is the effect of a marginal change in effort on the expected transfer received by the worker which unfolds via this very same performance measure. In other words, fix all but one signal of an incentive scheme and take the derivative of the transfer with respect to effort. The larger the absolute value of the derivative, the more important is this signal in determining the worker's wage in dependence of effort, and the stronger is its effect on effort.

Expected Performance Pay

Empirical data on incentive schemes is sparse, especially concerning contractual details such as the wage rate for a particular signal or the size of a bonus. Exceptions are few rather small self-collected data sets (Slade 1996, Gibbs et al. 2002). Even those data sets do not contain the relevant information to compute the marginal effect of effort on pay which unfolds via firm performance measures. Thus, a proxy for the extent to which firm performance measures are used in an incentive scheme is needed. A promising candidate is the expected total amount of pay which the worker receives due to firm performance measures. To justify this proxy, the importance of signals for the behaviour of the worker is revisited by means of an example.

Consider a bonus scheme based on a signal which takes on two values: high and low. Suppose that the employer pays the worker a base salary and a premium whenever the desired high outcome is realised, and nothing otherwise. Then, the worker's behaviour is driven the more by the signal, the higher the premium. Essentially, the expected difference in payoff between the low and the high outcome is the discrete analogue to the marginal effect on expected transfer from the definition above. Because the transfer for the low outcome is standardised to zero, the expected difference in pay is identical to the expected performance pay and amounts to the premium weighed by the probability of the high outcome to occur. Thus, there is a measure of the extent to which this dichotomous signal is used: the expected performance pay which the worker receives due to this signal.

The dichotomous nature as well as the feature that the low outcome does not result in a transfer are no particularities of this example; together with linear incentive schemes, discrete bonus schemes seem to be the prevalent form of incentive schemes (Slade 1996, Paarsch and Shearer 1999, Lazear 2000, Gibbs et al. 2002). In addition, fixed payments are usually declared as base salary and not assigned to the bad outcomes of performance measures, so that either the bonus is paid on top of the base salary when a certain goal was met or only the base salary is paid. This suggests that expected performance pay proxies the intensity of performance measures for many real-life instances.

In linear incentive schemes, expected performance pay can be related to the wage rate of the performance signal. If the wage rate is positive so that – similar to the bonus scheme – the agent is honoured for good outcomes rather than punished for bad outcomes, an increase in the wage rate implies an increase in expected performance pay. Unless the signal is scaled down, an increase in the wage rate implies steeper incentives and thus a higher effect of the respective signal on worker's behaviour. Hence, larger expected performance pay also indicates higher intensity of the performance measure in this case.

A third justification why performance pay proxies the intensity to which a performance measure is used is of psychological nature: It seems plausible that the worker devotes more attention and assigns more importance to signals which yield a higher expected payoff.

To summarise, firm performance pay seems to be a good proxy for the extent to which firm performance measures influence the behaviour of the agent.

3.3.3 Proxies for Misallocation Costs

When testing the hypothesis that large misallocation costs lead to intensive use of firm performance measures, one does not only need a proxy for the intensity of the use of firm performance measures, but also an indication for misallocation costs.

Unfortunately, misallocation costs themselves are not observable. Strictly following the definition, one would need a precise measurement of the benefit under two conditions: When effort allocation is balanced and under the second-best. While it is often already difficult to find a precise representation of the benefit because it might, for example, manifest itself partially in the future, one also has to deal with the problem of a missing counter-factual: It is necessary to operate the balanced incentive scheme under identical conditions, that is, at the same time with the same workers in the same plants, in order to generate the benefit resulting from balanced effort allocation. But there is an even more substantial problem: The balanced incentive scheme is a theoretical construct and need not exist in reality, in particular, because the implementation of any effort other than the second-best effort violates the assumption that employer and worker act rationally.

This should convince the reader that, while being an insightful theoretical concept, misallocation costs will hardly ever be measured in reality. On the other hand, this does not mean that it is impossible to find indicators for potentially large misallocation costs. Generally, it is clear that whenever there are multiple efforts, when the allocation between these efforts is not specified, and when employer and worker have different objectives, misallocation costs are present. More specifically, it was pointed out in the misallocation-uncertainty trade-off model in Section 3.2.2 and later in the discussion of the utilisation of firm performance measures in Section 3.3.1 that the sensitivity of the benefit function with respect to efforts explains when misallocation results in misallocation costs. If the worker's efforts are completely exchangeable, misallocation costs cannot arise. The same happens if the allocation of effort is regulated. Regulation is a useful tool to prevent misallocation, but it only works in standardised working environments. If the optimal effort allocation varies in unforeseeable ways, it becomes too costly to specify rules and monitor that the worker abides by them. When the worker has to respond to changes in his working environment in a particular way or when he is free to allocate effort and only certain effort combinations are beneficial, misallocation is likely to result in misallocation costs.

All these conditions relate to the job description, that is, the tasks the worker has to accomplish. Thus, task descriptors can be used to proxy for potential misallocation costs. The respective details are discussed in the subsequent empirical chapter.

3.4 Summary

Taking up a theme from the second chapter, the informational value of firm performance measures in incentive schemes can be explained by the misallocation they induce and the uncertainty they impose on the worker. Firm performance measures aggregate the contributions of individual workers; nevertheless, the idea that they lead to free-riding can be discarded on the basis that the theoretical evidence is not very compelling. Further, information about firm performance is relatively inapt to allocate workers to jobs because it does not vary between workers.

The main advantage of firm performance measures seems to lie in their potential to reduce misallocation when employed in an incentive scheme. On the other hand, they are likely to impose more uncertainty on the worker. It was shown in a misallocation-uncertainty trade-off model that a signal with these characteristics is more intensively employed when misallocation costs are large. Further, the degree to which efforts are exchangeable when generating benefit determines whether misallocation results in costs: The higher the substitutability, the lower are misallocation costs.

To test empirically whether the intensity of firm performance measures increases when misallocation costs are potentially large, the intensity can be proxied by firm performance pay. Because misallocation costs depend on the shape of the benefit function and because this shape also manifests itself in the tasks to be done at a job, task descriptors can indicate the need to reduce misallocation. In the next chapter, the empirical relationship between firm performance pay and task descriptors is analysed to test the hypothesis that firm performance measures are employed to reduce misallocation costs.

4

Statistical Model and Empirical Evidence

Building on the theoretical insights of the second chapter and their application to firm performance measures in the third chapter, this chapter examines the empirical evidence for the hypothesis that the value of firm performance measures results from the reduction of misallocation costs. Empirically, firm performance pay should be high whenever task descriptors indicate high distortion costs.

There are several empirical studies which implicitly analyse the value of signals because they deal with the question whether incentive schemes have beneficial effects or what type of incentive schemes are employed under which conditions. These studies are reviewed in Section 4.1.

In contrast to its predecessors, the present empirical study analyses the relationship between task descriptors and firm performance pay and for the first time uses data on individual jobs covering the private sector of a whole economy. The unique data source as well as its performance pay variables and task descriptors is presented in Section 4.2. Further, the empirical implications of the theoretical chapters are rephrased in terms of the available variables, and a censoring problem which affects the consistency of estimations is described.

To solve the censoring problem, an estimator that can be applied to the context of multivariate censoring is derived in Section 4.3. It applies to a considerably more general class of censoring problems, and it is shown to have the desirable properties of a maximum-likelihood estimator under similar regularity conditions.

The derived estimator is used in Section 4.4 to estimate the relationship between task characteristics and firm performance pay. The final section summarises this chapter.

4.1 Empirical Studies of the Value of Signals

The empirical literature which studies the use of signals in contracts and its effects operates under the title "incentive provision." Gibbons (1996) and Prendergast (1999) review a whole collection of empirical studies which deal with contract design and its consequences. Following Prendergast (1999), these studies can be classified into two groups. The first group is concerned with the question of whether incentives have an effect. It will be discussed in Section 4.1.1. The second group analyses whether employers choose incentive schemes according to the theory and is summarised in Section 4.1.2

4.1.1 Do Incentives Matter?

The overwhelming majority of empirical articles written on incentive schemes is concerned with the question whether incentive schemes have an effect on effort provision or not.

One of the most prominent examples is the analysis of Lazear (2000) who looks at the introduction of a piece rate scheme at Safelite Glass Corporation. When workers were paid by the number of installed auto windshields, this number rose significantly; the largest estimate for the increase amounts to 44%. Lazear attributes the effect partly to turnover in the work force and partly to increased effort. Numerous other examples showing the effect of incentives can be found in the overview articles of Gibbons (1996) and Prendergast (1999).

Generally, the literature on the effect of incentives is plagued by the problem that the choice of the incentive scheme is endogenous. Managers are more likely to introduce incentive schemes if performance needs to be increased and the incentive scheme is expected to achieve this goal. If this expectation is on average positively correlated with the true effect, the incentive effect is overestimated because all observed incentive schemes operate under conditions which are favourable to them. In Lazear's windshield example, piece rates were introduced in the windshield installation department, and this department was probably selected because the management expected to achieve higher beneficial effects there than in other departments.

A very similar problem occurs when the impact of interventions is evaluated. In particular, the *evaluation of labour market programmes* suffers from the possibility that from the pool of eligible participants, the more capable ones might self-select into these programmes, so that their effect is overestimated. In biometrical terms, participation in the programme is the *treatment* which is administered to the participant. Similarly, the incentive scheme can be regarded as a treatment affecting some workers but not others. In both cases, the treatment choice is not controlled by the econometrician. Hence, comparing the treated and the untreated does not yield the pure treatment effect.

There are several methods of dealing with the problem of an uncontrollable treatment effect (Heckman and Robb 1985). While these methods have been widely applied in the evaluation of labour market programmes, they have not yet gained ground in the incentive effect literature. One reason may be that the application of many of these methods is not readily transferable to the incentive effect problem because it is not the selection of the observational units into the treatment which poses the problem but the selection of the treatment itself. In other words, the problem does not result from the workers selecting into jobs but from the choice of the incentive scheme. To avoid biased estimates, instrumental variables or structural models could be used. Both methods require that the choice itself has been carefully studied. This outlines the importance of studies which –like this book– are concerned with modelling the choice of incentive schemes.

The endogenous character of the choice of incentive schemes is explicitly taken into account in the work by Shearer and Paarsch (1999, 2000). They analyse the incentive effect of piece-rates at a British Columbia tree-planting firm. The comparison of workers receiving piece-rates with workers who get a fixed salary is taken as an upper bound to the productivity increase, because piece-rates are chosen under planting conditions which are to their advantage. As a lower bound, they consider the average wage difference between the two worker groups. They argue that the employer would prefer fixed pay to piece rates if the productivity gain by piece rates does not compensate for the additional wage costs. In addition, Shearer and Paarsch model the choice of fixed pay versus piece-rate and estimate a structural model to derive the incentive effect of piece-rates. In another study, Shearer (2000) uses data from a social experiment conducted at the same tree-planting firm. To generate this data, the type of contract was randomly allocated to workers. The difference between treated and untreated workers reflects the true incentive effect, and is not blurred by the endogenous choice of the contract.

Incentive schemes are employed under varying conditions; not all schemes are equally suited to these different conditions and some are especially designed for specific situations. Prendergast (1999) is aware that his question "Do incentives matter?" cannot be answered generally because the effect of incentives depends on the conditions under which they operate. The answer can only be a collection of case studies which show that there are conditions for which an appropriately crafted incentive scheme induces workers to alter their behaviour. The degree to which the behaviour is altered, the incentive effect, is not overestimated when it is applied to the original context. Only if the incentive effect is claimed to have the same magnitude independently from the conditions – for example in different departments, firms, or industries – the selection critique has bite. Then, the magnitude is likely to be overstated.

The next section is concerned with empirical studies which exploit the variation in the production context to explain variation in the incentive scheme.

4.1.2 Choice of Incentive Schemes

Only a small fraction of the literature addresses the employer's choice of incentive schemes. One reason might be the lack of detailed data about incentive contracts. To be observable to the econometrician, an incentive contract has to link payments explicitly to performance measures. Informal agreements cannot be directly measured. But a large share of real-life incentive schemes is based on the implicit understanding between employer and worker that the latter will be rewarded when acting appropriately. These contracts can never be directly observed. Being unable to observe the complete scheme of incentive provision, research has focused on particular observable aspects and only one attempt has been made to analyse the incentive provision as a whole (MacLeod and Parent 1998).

Ferrall and Shearer (1999) study the one-dimensional trade-off between uncertainty and incentives, which is intertwined with a free-rider issue resulting from the fact that team output is equally shared. They use payroll records from the Britannia Mining and Smelting Company from the 1920s and find that about half of the difference between the first-best and the actual surplus can be attributed to unsophisticated design of the incentive scheme. The other half are non-contractibility costs. According to them, Britannia could have almost doubled its expected profit using the optimal second-best contract, a highly non-linear scheme.

While the study of Ferrall and Shearer (1999) compares the efficiency of different incentive schemes under identical conditions, Slade (1996), Dupuy and Lafranchi (1998), MacLeod and Parent (1999), and the empirical analysis pursued in this monograph study the variation in incentive schemes and relate it to variations in the the environment under which workers operate.

The most comprehensive study of the variation of incentive schemes in response to external conditions has been put forward by MacLeod and Parent (1998, 1999). They distinguish the coarse classes of input- and output-based compensation and break down these classes further until they end up with five types of incentives: commission rates, salary plus bonus pay/promotion, salary plus termination, hourly wage, and piece rate. This classification can be depicted as a tree of sequential decisions and MacLeod and Parent (1998) suppose that the employer determines the incentive scheme following this sequence. Using various data sets (Current Population Survey, Panel Study of Income Dynamics, National Longitudinal Survey of Youth, and the Quality of Employment Survey), they sequentially estimate the dichotomous decisions using linear probability models. Because there is no information for both environmental conditions and the choice of incentive scheme on an individual level, they compute work characteristics within occupational cells and match them to incentive choices in respective cells. With respect to multitasking, they find that the number of tasks is associated with more incomplete contracting. Firm performance pay is not included in the data set. Hence, the relationship between firm performance pay and multitasking is not examined by MacLeod and Parent. This relationship is the focus of the subsequent empirical analysis.

Dupuy and Lafranchi (1998) take a less general approach than MacLeod and Parent (1999) and focus on the question of whether to use absolute or relative performance measures. A major advantage of relative performance evaluation is that common shocks affecting all workers can be filtered out so that the income uncertainty can be substantially reduced; theoretical models which formalise this idea were developed by Nalebuff and Stiglitz (1983) and Green and Stokey (1983). On the other hand, relative performance measures might induce destructive behaviour and sabotage activities (see, for example, Lazear 1989). Dupuy and Lafranchi (1998) use the same data set on French private firms which will be exploited later in the empirical analysis. However, they analyse information from the firm level while individual level data is examined here. They use a multinomial logit model to regress the three possible incentive choices (no bonus scheme, absolute performance evaluation, relative performance evaluation possibly in addition to absolute performance evaluation) on environmental variables. In accordance with the theory, they find that volatile market conditions lead to a stronger incidence of relative performance measures.

The work of Slade (1996) comes closest to the present analysis. Slade investigates the relationship between multitasking and contractual choice using the data from petrol stations in Vancouver. There are various ways in which an oil company and the operator of a station can design their relationship, ranging from the oil company being the owner of the station and the operator being a salaried employee to commission contracts, leasing contracts, and operator-owned stations. This contractual variation can be related to secondary services offered by the station beyond the selling of petrol. For example, the station can include a convenience store or a repair shop. In both cases, the revenue from the second activity usually accrues to the operator of the station. Slade argues that complementarities in demand between petrol and convenience-store items are larger than those between patrol and repair services. As a consequence, operators whose secondary activity is a repair shop need steeper incentives, while operators whose secondary activity is a convenience-store are to some degree already motivated: Increasing petrol sales raises their revenues from the convenience-store. Slade estimates a sequence of probit models, and the data confirms her hypothesis.

Similarly to Slade (1996), the present work examines the variation in the characteristics of a multitask setting and the resulting changes in the contractual design. In contrast to her analysis, the data here covers several industries and occupations. Accordingly, information as well as predictions about the contractual choice will be less specific.

4.2 The ECMOSS Data Set

In Section 3.3, the foundations for the empirical model were laid: The hypothesis that firm performance measures are used more extensively to avoid misallocation costs can be examined using an an empirical model which relates firm performance pay to task descriptors. As already seen in the discussion of other empirical studies, data sets which contain information on contractual features and task characteristics are rare. To overcome this problem, Slade (1996) and Gibbs, Merchant, der Stede, and Vargus (2002) collect their own data, and MacLeod and Parent (1999) link different data sources on the basis of occupational categories. As a consequence, empirical findings reflect either very specific settings in case of the self-collected data sets, or they rest on the validity of the link between data sets. In contrast, the data set which is presented in this section reflects a broad spectre of production settings and contains information on contractual features and task descriptors for the same individual.

4.2.1 Survey Design

The French National Institute of Statistics and Economic Research (Institut National de la Statistique et des Etudes Economiques - INSEE) regularly conducts a labour cost survey. In 1992, this survey was supplemented by detailed questions on the salary structure. The cross-sectional data set for 1992 which combines the labour cost survey and the information on the salary structure, the Enquête sur la Coût de la Main-d'Oeuvre et la Structure des Salaires (ECMOSS), provides a good data base to study incentive issues on a broad basis; it is designed to be representative for firms with more than 25 employees in the private non-agricultural sector of France. Because ECMOSS is an official survey, answering was mandatory.

The survey consists of three parts (see also Figure 4.1). For the first part, about 15,000 establishments were asked to give information about general labour costs and remuneration policies on the establishment level. Second, the establishments were required to supply information on the size and composition of salaries for a sample of workers from the personnel files. This led to around 150,000 worker observations.

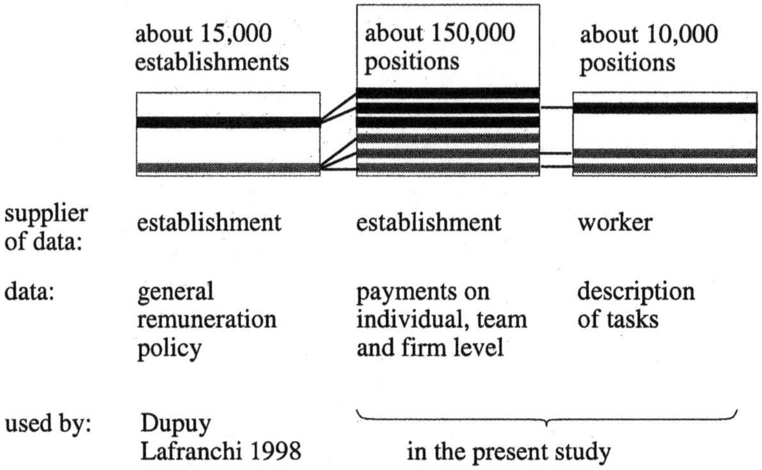

Fig. 4.1. Survey design of the ECMOSS data set

Finally, questionnaires were sent to a subsample of about 10,000 of these workers. Identifiers allow us to link the individual information from the establishment to that from the worker.

4.2.2 Performance Pay Variables

The data set does not contain detailed information about contracts; this information would probably have been too complex to survey and to handle considering the large variety of contractual arrangements which exist in the private sector of a whole economy. However, information is provided on payment supplements such as compensation for difficult working conditions, individual performance pay, team performance pay, firm performance pay, and others. In principle and as argued in the previous chapter, firm performance pay can be used to proxy the degree to which firm performance measures are utilised. But firm performance pay will not be observed under all circumstances: Even though the worker received performance pay it is only recorded if it belongs to the five largest payment supplements. This censoring problem will be extensively discussed in the next section. Due to this problem, the frequency of observed payments in the data set which is depicted in Table 4.1 is only a lower bound for the incidence of performance pay. This table shows that around two-thirds of the working population in the non-agricultural private sector received no performance pay. French enterprises seem to be much more inclined to have a firm performance pay scheme than their German counterparts: About 37% of the establishments in the ECMOSS data set offer such schemes in comparison with only 11% in Germany in the same year (Pannenberg and Schröder 2000).

The difference in the usage of firm performance pay between France and Germany may result from the longer tradition of firm performance pay in France. The first legal framework to induce firm performance pay was already devised in 1959. In 1967, it became obligatory for French firms with more than 100 employees to let workers participate in their financial success. In 1986, the Ordonnance Rélative à l'Interessement et la Participation des Salariés aux Résultats de l'Entreprise (no 86-1134) was passed. It lowered the threshold from 100 to 50 employees and defines the legal situation at the time of the survey in 1992. This ordonnance requires firms to set up a fund, the Réserve Spéciale de Participation. The size of this fund is a prescribed function which depends on the after-tax surplus, firm's capital stock, the total sum of salaries, and the value-added. The fund must be distributed to workers either equally or proportionally to workers' wages or hours or a weighted mean of both.

Besides the obligatory firm performance pay, different voluntary forms of firm performance pay exist; one of these forms, the so-called *interessement*, is also legally covered by the ordonnance mentioned above. It should be emphasised that the variable firm performance pay which is used in the present analysis only encompasses voluntary forms of firm performance pay (amongst them the *interessement*) and not the obligatory form (*participation*).

The information concerning the question whether firm performance pay occurred and whether this occurrence is related to the incidence of team and individual performance pay has been analysed by Schnedler (2000, 2001) using multinomial logit and multivariate probit models. It turns out that the incidence of firm performance pay is positively associated with more responsibility and professional social interaction on behalf of the workers. In addition, the occurrence of firm performance pay is significantly negatively related to the occurrence of individual performance pay, which suggests that the two payment methods are substitutes. In contrast to these studies, the following analysis is concerned with the actual level of firm performance pay. Therefore, it uses infor-

Table 4.1. Frequency of observed payments

Type of performance pay	Observations	Frequency
individual	16100	10.9 %
only team	4177	2.8 %
only firm	22097	14.9 %
individual and team	881	0.6 %
individual and firm	3401	2.3 %
team and firm	1293	0.9 %
individual, team and firm	1490	0.1 %
total	148258	100 %

The depicted observed frequencies for performance pay can be regarded as a lower bound for the use of the respective performance measures (Source: own calculations using ECMOSS 1992).

mation previously discarded. On the other hand, the focus is on firm performance pay only and not on individual or team bonuses.

4.2.3 Task Descriptors

The task environment of workers is captured by several variables. Workers are asked whether they have subordinates (MANAGE), about the number of subordinates (SUBORDx), whether they have to carry out precisely describable tasks (EXECUTE), solve problems without consulting their supervisor (RESPONS), and have non-hierarchical professional contact to other departments (COOPERATE).

The number of subordinates was recoded into four categories to allow for non-linearities; the variables SUBORD1 to SUBORD3 indicate the category: none, one to five (SUBORD1), six to fifteen (SUBORD2), and more than fifteen employees (SUBORD3). The variables EXECUTE and RESPONS are dichotomous, the variable COOPERATE was recoded to be dichotomous. This was done to eliminate subjective assessment of the word "sometimes", assuming that the distinction between professional contacts and no such contacts is less ambiguous. For the precise wording of questions and coding see Appendix C.

All variables other than EXECUTE indicate high potential misallocation costs: Having subordinates involves several tasks of differing importance and a non-controllable allocation of effort between tasks. These positions are thus more vulnerable to misallocation in comparison to other jobs. Autonomous problem solving (RESPONS) presumes a thinking process which is difficult to control. In addition, the quality of this process can hardly be individually measured without the reference of a correct decision. Having such a reference would imply knowledge of the correct decision, in which case it is not necessary to hire a worker to find a decision to begin with. In absence of a performance measure for the global objective, it is thus not possible to align the worker's interest with the employer's objective, and misallocation costs are likely to be higher in comparison with a worker who need not decide. Non-hierarchical professional contacts with other departments (COOPERATE) indicate that there is a need for coordination or cooperation amongst workers. Even if the worker's input to the meeting consists solely of providing information on which others base their decision, the worker has influence on this decision by concealing or revealing certain information for strategic purposes, or by acquiring not sufficient information (Aghion and Tirole 1997). Hence, given the need for non-hierarchical contacts, the potential for misallocation is likely to be higher. Finally, if there is a precise description of tasks to be executed (EXECUTE), there is less autonomy for the worker, less room for misallocation and therefore misallocation costs are lower.

4.2.4 Expected Relationships Between Tasks and Pay

The preceding section predicts that expected firm performance pay should be higher when workers have subordinates (SUBORDx), solve problems autonomously (RESPONS), and when they have non-hierarchical contacts (COOPERATE). Firm performance pay is expected to be reduced if tasks can be clearly separated and described (EXECUTE).

Suppose that firm performance pay is not observed if it is below a censoring threshold. As explained later, this does not imply that no firm performance pay occurred. How should we treat such an observation? If we assume that unobserved firm performance pay is zero, and then compute the overall average, we obtain a lower bound for the true average of firm performance pay. Likewise, if we assume the unobserved firm performance pay to be equal to the value of the censoring threshold, this yields an upper bound on the true average. Table 4.2 shows the lower and upper bound for the average firm performance pay and relates them to different tasks. Lower and upper bounds do not overlap for any task, so that average firm performance is unambiguously rising or falling. The observed effects accord with the predicted effects: Average firm performance pay is larger for workers with at least one subordinate (MANAGE), for workers who solve problems autonomously (RESPONS), and for workers with non-hierarchical professional contacts (COOPERATE) and smaller when the worker's tasks can be clearly described (EXECUTE). Therefore and in spite of the censoring problem, the data confirms the predicted relationship between tasks and firm performance pay on this purely descriptive level.

Table 4.2. Average firm performance pay and tasks

	Average firm performance pay			
Task	no		yes	
	lower bound	upper bound	lower bound	upper bound
MANAGE	1273	1436	3553	3744
JUDGE	1718	1888	3022	3256
EVALUATE	1493	1662	5417	5667
RESPONS	772	986	2360	2536
COOPERATE	970	1095	2073	2278
EXECUTE	2597	2800	1391	1566

Certainly there are other effects unrelated to misallocation which may induce a correlation between firm performance pay and task descriptors. For example, workers who tend to be less risk-averse might prefer jobs which include those tasks that indicate high potential misallocation costs. Then, firm performance pay would be higher for these jobs, independently of misallocation considerations and only because uncertainty compensation is lower. This effect is difficult to control for since risk-aversion is not directly observed. Nevertheless, risk-aversion is closely associated to individual variables. Analysing the Brabant survey, Hartog and Jonker (1999) found that risk-aversion drops with schooling level, income, and wealth and is larger for females. Hence, these variables can be used as proxies. Hartog and Jonker (1999) cannot examine the effect of age because all individuals in the Brabant survey are of the same age. Nevertheless, age seems to be an important determinant of the attitudes towards uncertainty, and thus it is also sensible to control for age.

Another explanation for the relationship between tasks and performance pay could be the hierarchical rank. Workers at high-rank positions are likely to have more influence on the overall firm performance. Thus, the ratio of the variation of the performance measure under control versus the variation beyond the control of the worker *(signal-noise ratio)* is larger, and the use of the firm performance measure becomes relatively cheaper. At the same time, higher ranks may be associated with particular tasks, so that a significant relationship between tasks and firm performance pay is observed when rank is ignored. Unfortunately, there is no direct measure of rank in the data set; however, because hierarchical rank is very closely related to base pay, a good proxy is available. In addition, we have to be careful when interpreting the number of supervised employees (SUBORDx) because this number is also related to rank.

Finally, there could be effects of firm characteristics such as the product or service produced, the market conditions, or the firm's culture, which on the one hand lead to firm performance pay and on the other result in particular tasks. For example, the firm's product might allow a precise measuring of firm performance, so that firm performance pay becomes attractive while it also requires a particular production technology which entails specific tasks. Fixed firm effects will capture these phenomena.

Descriptive statistics for the control variables can be found in Table C.2.

4.2.5 The Bonus-pay Censoring Problem

It has already been mentioned that the observed performance pay is characterised by a censoring problem. The nature of this problem results from the questionnaire design. Bonuses to the worker were classified into eight categories, those linked to

- base pay (e.g. holiday, 13th month income)
- personal characteristics (e.g. seniority, marriage, birth)
- special events (e.g. firm's anniversary)
- workplace constraints (e.g. cleanliness, expatriation)
- individual performance (e.g. efficiency, exceptional service, innovations)
- team performance (e.g. meeting objectives, joint service)
- firm performance (e.g. balance sheet, results, profit), and
- legal circumstances (e.g. transportation compensation).

The titles of the categories and the examples are translations from the original questionnaire. Firms were asked to name the five largest categories for each individual worker and to state the respective amount of French Francs (FRF) paid and the sum of the three smallest categories. For an illustration of the information situation, consider Figure 4.2. In this example, the worker receives seven types of bonus payment. In particular, the worker gets firm performance pay. Since there are five types for which the payment is larger, firm performance pay only shows in the sum of the unobserved components.

supplement type	true values	observed labels	observed values
fixed to base pay	15,000 FRF	fixed to base pay	15,000 FRF
personal bonus	0 FRF	condition compensation	8,000 FRF
condition compensation	8,000 FRF	legal compensations	7,000 FRF
firm performance bonus	4,000 FRF	indiv. perform. bonus	6,000 FRF
team performance bonus	5,000 FRF	team performance bonus	5,000 FRF
indiv. performance bonus	6,000 FRF		
event bonus	2,000 FRF	sum of unobserved suppl.	6,000 FRF
legal compensations	7,000 FRF		

Fig. 4.2. Censoring example

Although the worker receives firm performance pay, it is not observable; five other types of supplement are larger.

The sum of the unobserved components works as an indicator for potential censoring of firm performance pay. In about 85% of the cases, firm performance pay is not observed, but the sum of the unobserved components is larger than zero. Depending on whether one sets the value for the unobserved firm performance pay to zero or to the smallest observed value, the estimated average firm performance pay varies substantially; the lower and upper bounds in Table 4.2 provide a good example. All this suggests that censoring might be a severe problem and that consistent estimation of the relationship between task characteristics and firm performance pay has to take censoring into account.

4.3 Estimation Under Multivariate Censoring

This section deals with the problem of consistent estimation when multivariate observations are censored. Section 4.3.1 briefly reviews estimation techniques and explains why they cannot be used for the bonus pay censoring problem at hand. Section 4.3.2 presents a rule to construct consistent estimators for a large class of censoring problems. Finally, this rule is applied to the bonus pay censoring problem in Section 4.3.3.

4.3.1 Survey of Estimation Methods

Tobin (1958) was the first to point out to a broad econometric audience that estimating a linear model by ordinary least squares is problematic when the dependent variable can only take positive values. He also suggested an alternative, the – now famous – tobit estimator. The econometric community had to wait another 15 years before Amemiya (1973) proved that Tobin's intuition was correct and that his proposal yields a consistent and root-n asymptotic efficient estimator. In the following years, a large variety of censoring models and estimators were developed. Amemiya (1984) surveys this literature and classifies models into five types. All five types build on normally distributed error terms and look at fixed (possibly unknown) thresholds above or below which certain variables are observed. Since the bonus pay censoring problem has a varying rather than a fixed threshold, they cannot be used here. Nelson (1977) proposes an estimator for varying thresholds, but it only applies to a single censored variable, and he neither proves its consistency nor its asymptotic distributional properties.

A robust alternative to these estimators when there is censoring is the least absolute deviation estimator. Fitzenberger (1997) provides an overview on this estimation technique. The effect of censoring can be reduced or entirely avoided because the mean which is vulnerable to censoring is replaced by the median, a robust location estimator. However, the sample median must be observed, otherwise the estimator generates no meaningful results. Since firm performance pay is observed in less than 18 % of the cases (see Table 4.1), these methods cannot be applied to the bonus-pay censoring problem.

For many economic problems with censored variables, estimation follows the general design of Amemiya (1984): Contributions for censored and uncensored observations are derived and the resulting function is maximised. Often the respective estimator is called maximum-likelihood estimator where in fact it does not maximise a likelihood but a mixture of a density and a probability function. It is then assumed that the result has the desirable properties of a regular maximum-likelihood estimator such as consistency and root-n asymptotic normality. But with the exception of Amemiya (1973), there is no proof which justifies this assumption. This is problematic because Amemiya's proof only concerns the rather restrictive case of a one-dimensional normally distributed random variable which is censored at a fixed threshold. In addition, no explicit rule exists as to how the contribution of a censored observation has to be calculated in order to get a consistent and root-n asymptotically normally distributed estimator. Subsequently, such a rule is provided for a general class of censoring problems. Then, the rule is applied to the bonus pay censoring problem to yield an estimator.

Like other estimators for censoring problems, the estimator based on this rule maximises a mixture of density and probability contributions. Thus, it is not a maximum-likelihood. However, it is shown to have the properties of a maximum-likelihood estimator. Note that, although the objective function is not a likelihood function, the resulting estimator is not a *pseudo-maximum likelihood estimator* (Gouriéroux, Monfort, and Trognon 1984), either; such estimators are obtained when the true density is different from the assumed density. Here, like in maximum likelihood techniques, the estimator is based on the correct density.

4.3.2 Constructing an Estimator

The starting point of the construction is an arbitrary censoring problem. This means that the interest lies in estimating the p-dimensional parameter $\theta \in \mathbb{R}^p$, which governs a continuously distributed random vector $Y = (Y_1, \ldots, Y_q)$ with a joint density function $f(y, \theta)$. Let $y = (y_1, \ldots, y_q)$ be a particular realisation of this random vector. As censoring is the focus of attention, not all components of this realisation will always be observed, and whether a particular component y_j is observed will depend on the realisation y. Formally, define:

Definition 1 (visibility). *The set* $\mathcal{V}_j := \{y | y_j \text{ observed}\}$ *is called the* visibility set *of* j. *The complement to the visibility set is denoted by* $\bar{\mathcal{V}}_j$. *The function*

$$\begin{aligned} v_j : \mathbb{R}^q &\to \{0, 1\} \\ y &\mapsto v_j(y) = 1\!\!I_{\mathcal{V}_j}(y) \end{aligned}$$

is called visibility function *of* j. *The random variable* $V_j = 1\!\!I_{\mathcal{V}_j}(Y)$ *is called* random visibility *and* $v_j = 1\!\!I_{\mathcal{V}_j}(y)$ *the* (realised) visibility *of* j.

If the observation y is an element of the complement of the visibility set, $y \in \bar{\mathcal{V}}_j$, then the j-th component of the vector is not visible ($v_j =$

0). Since probabilities should be attached to the event that the j-th component is visible, the following assumption is necessary:

Assumption 1 *For all j, the visibility set $\mathcal{V}_j$ is (Lebesgue-)measurable.*

This is the only restriction imposed on censoring conditions which is substantially more general than the restriction put forward by Amemiya (1984), who supposes that $\mathcal{V}_j$ is a one-dimensional interval.

Since $\mathcal{V}_j$ is measurable, the probability of observing j is:

$$P(v_j = 1) = \int \mathbb{I}_{\mathcal{V}_j}(y) f(y, \theta) dy = \int_{\mathcal{V}_j} f(y, \theta) dy.$$

Collecting the visibilities of all components of Y in a vector, one gets the random vector $V = (V_1, \ldots, V_q)$, the respective realisation is $v = (v_1, \ldots, v_q)$. Since any component of V can only take two values, there are 2^q different realisations of V. These realisations can be numbered from $s = 0$ to $s = 2^q - 1$ and a new random variable S can be introduced which denotes the *state of visibility*. Each realisation s of S can be associated with a vector $v^s = (v_1^s, \ldots, v_q^s)$ which indicates the visible components in state s. The label $s = 0$ is used for the the particular state in which no component is visible, $v^0 = (0, \ldots, 0)$.

Which realisations y of Y lead to the realisation s of S? In particular, what is the set $\mathcal{V}^s$ such that $y \in \mathcal{V}^s$ results in the visibility state v^s? This set must be composed such that y is in $\mathcal{V}_j$ for those components j which are visible ($v_j^s = 1$) and in $\bar{\mathcal{V}}_j$ for those components j which are not visible ($v_j^s = 0$) in state s. Formally,

$$\mathcal{V}^s = \bigcap_{\{j|v_j^s=1\}} \mathcal{V}_j \cap \bigcap_{\{j|v_j^s=0\}} \bar{\mathcal{V}}_j.$$

The set $\mathcal{V}^s$ is (Lebesgue-)measurable because it is defined as the intersection of (Lebesgue-)measurable sets. Thus, it becomes possible to compute the probability of a state s to occur:

$$P(S = s) = P(y \in \mathcal{V}^s) = \int_{\mathcal{V}^s} f(y, \theta) dy.$$

Any estimator has to be defined in terms of observables. To generate observables from the partly unobserved realisation y, it is necessary to introduce an operator which extracts the observed components of y. Also, it is necessary to select the unobserved components, for example to indicate which variables need to be integrated out.

Definition 2. *Denote by ν^s an operator which extracts the visible components of y:*

$$\begin{aligned}\nu^s : \quad & \mathbb{N} \times \mathbb{R}^q \rightarrow \mathbb{R}^{l(s)} \subseteq \mathbb{R}^q \\ & (s, y) \mapsto (y_{j1}, y_{j2}, \ldots, y_{j_{l(s)}}),\end{aligned}$$

where $j_1, \ldots j_{l(s)}, \in \{j | v_j^s = 1\}$ and $l(s)$ is the number of observed components in state s. Define $\bar{\nu}^s$ to be an operator which extracts the unobserved components:

$$\begin{aligned}\bar{\nu}^s : \quad & \mathbb{N} \times \mathbb{R}^q \rightarrow \mathbb{R}^{q-l(s)} \subseteq \mathbb{R}^q \\ & (s, y) \mapsto (y_{j1}, y_{j2}, \ldots, y_{j_{q-l(s)}}),\end{aligned}$$

where $j_1, \ldots, j_{l(s)} \in \{j | v_j^s = 0\}$.

To see how the operators work, reconsider the example from Figure 4.2. The vector y is the vector of pay supplements:

$$y = (y_{\text{fix}}, y_{\text{personal}}, y_{\text{comp}}, y_{\text{firm}}, y_{\text{team}}, y_{\text{ind}}, y_{\text{event}}, y_{\text{legal}}).$$

In the example, neither personal bonus nor firm performance bonus or event bonus were observable. Respectively, the visibility is: $v^s = (1, 0, 1, 0, 1, 1, 0, 1)$. Now, ν^s selects all visible supplements, so that

$$\nu^s y = (y_{\text{fix}}, y_{\text{comp}}, y_{\text{team}}, y_{\text{ind}}, y_{\text{legal}})$$

while its counterpart selects the non-visible components:

$$\bar{\nu}^s y = (y_{\text{personal}}, y_{\text{firm}}, y_{\text{event}}).$$

The visibility operators can be used to define a contribution when the state is s and the observation is $\nu^s y$. Consider the function:

$$\tilde{f}^s(\nu^s y, \theta) := \int_{\mathcal{V}^s} f(y, \theta)\, d(\bar{\nu}^s y),$$

The function only depends on observables and is well-defined because $\mathcal{V}^s$ is (Lebesgue-)measurable. However, it is not a density as integration over the remaining visible arguments, $\nu^s y$, does not yield one but the probability of the state s. Nevertheless, the function has the same arguments as the conditional density given the state s, $f^s(\nu^s y, \theta)$, and is proportional to it:

$$\tilde{f}^s(\nu^s y, \theta) = P(S = s) \cdot f^s(\nu^s y, \theta).$$

Suppose there are $i = 1, \ldots, n$ independently drawn observational units, so that $y^i = (y^i_1, \ldots, y^i_q)$ is the realisation of the random vector for the observational unit i and s_i the respective state. A first approach to estimate the parameter θ could be to use the product of the conditional densities $f^{s_i}(v^{s_i}y^i, \theta)$ for all states where at least one variable is observed and maximise it in the parameter. Since the conditional density gives an accurate description of the observed distribution of y, the unknown parameter can be recovered consistently. However, the estimator discards the information which is conveyed in the case that no variable is observed. Hence, it cannot be asymptotically efficient. In addition, it involves computing the probability of every state s to occur.

Alternatively, take $\tilde{f}^{s_i}(v^{s_i}y^i, \theta)$ as the contribution of observation i, and consider the following estimator:

$$\hat{\theta}_n := \text{argmax}_\theta \prod_{i=1}^{n} \tilde{f}^{s_i}(v^{s_i}y^i, \theta). \tag{4.1}$$

In contrast to the estimator using conditional densities, there is a well-defined contribution when no variable is observed ($s = 0$): $\tilde{f}^0(\cdot, \theta)$ is precisely the probability of this to occur, $P(S = 0)$. Thus, the information which is contained in not observing y is not discarded and the estimator may be asymptotically efficient. The estimator has a second interesting feature: It embeds the class of tobit type I estimators introduced by Amemiya (1984) as a special case which can be obtained by considering a single variable ($q = 1$) and two states, one in which the variable is observed ($s = 1$) and one in which it is not ($s = 0$).

Besides the tobit type I estimator, the proposed modelling framework embeds many other censoring models: the censoring which leads to the tobit type III and type IV estimator in the classification of Amemiya (1984)[1], the model considered by Rosett and Nelson (1975), and the model of Nelson (1977). As already mentioned, these and many other studies, which have to tackle censoring issues, use "likelihood" functions which are neither density nor probability functions but rather mix continuous and discrete contributions. However, Davidson and MacKinnon (1993, p. 539) point out: "Because of this mixture of discrete and continuous random variables, standard proofs for the consistency and asymptotic normality do not apply [...]." While it is certainly possible to write down an objective function, calling it a "likelihood" does

[1] The approach can also accommodate type II and type V from Amemiya's typology but the respective extension is not straight forward.

not imply that the maximiser of this function is consistent and root-n asymptotically normal. Only for the particular case that Y is a univariate normally distributed variable and censored below a fixed threshold does the proof of Amemiya (1973) ensure that the maximiser of the mixture of continuous and discrete contributions indeed has the properties of a regular maximum likelihood estimator.

As can be seen from Appendix A, the results of Amemiya (1973) generalise under mild conditions to the estimator defined by (4.1). Namely, this estimator is consistent and the square root of the number of observations n times the estimator, the *root-n estimator*, is asymptotically normally distributed and asymptotically efficient. In a sense, this *ex post* justifies the use of "likelihood" functions, which are composed of continuous and discrete distributions if they can be written in the form of (4.1) and if the additional conditions are fulfilled. It also provides a flexible framework to model censoring problems and permits consistent estimation in this framework.

4.3.3 Application to the Bonus Pay Censoring Problem

Formula (4.1) should now be used to derive a consistent estimator for the relation between the censored firm performance pay and the task descriptors in the ECMOSS data set. The task descriptors together with control variables are collected in a matrix X, and the eight bonus categories can be identified as components of Y. The unknown parameter θ_j is taken to be the uncensored expected pay of the j-th component, so that θ_{firm} is the uncensored expected firm performance pay. However, it should be allowed that this parameter varies in the realisation x^i of X task descriptors and control variables of the i-th worker:

$$\theta_j^i = x^i \beta_j,$$

where β_j is a parameter vector that describes the relation between the worker's variables and the expected level of the j-th bonus component; the formalities related to this parameterisation are discussed in Appendix A.4.

Since only the five largest components of the wage components are observed, the visibility sets take the form $\mathcal{V}^j = \{y_j^i \geq y_{(5)}^i\}$ for the i-th observation where $y_{(5)}^i$ denotes the fifth-largest pay supplement observed for worker i.

Although the sum of the unobserved components is available, this information is ignored here. Doing so avoids the convolution of the unobserved variables and the calculation of the estimator simplifies. However, this simplification comes at a price: The root-n-estimator is not asymptotically efficient anymore. As information is discarded, the variance of the root-n-estimator exceeds the lower bound. Note that ignoring the sum of unobserved components does not affect consistency.

Using the visibility sets, the contribution $(\nu^{s_i}y^i, x^i)$ of the i-th observation, becomes:

$$\tilde{f}^{s_i}(\nu^{s_i}y^i|\beta, x^i) = \int_{-\infty}^{y^i_{(5)}} \int_{-\infty}^{y^i_{(5)}} \int_{-\infty}^{y^i_{(5)}} f(y^i|\beta, x^i)\, d(\bar{\nu}^{s_i}y^i).$$

Even without the convolution, there are still three integrals which need to be solved. If for example $f(\cdot)$ is the normal density, this implies a three-dimensional numerical integration. Alternatively, one could employ the technique of Gouriéroux and Monfort (1996) and simulate the three-fold integral. To do so, one draws realisations from the unobserved components, considers only those draws where the realisations fall below $y^i_{(5)}$, and computes the average over these draws which – by the law of large numbers – is a proxy for the integral.

Here, a different approach is taken. It is assumed that the bonus components Y are independent given X and β. Then, the density can be decomposed: $f(y|\beta, x^i) = \prod_j f_j(y_j|\beta_j, x^i)$. This allows to simplify the contribution of i to:

$$\tilde{f}^{s_i}(\nu^{s_i}y|\beta, x^i) = \prod_{\{j|j \text{ not visible}\}} F_j(y^i_{(5)}|\beta_j, x^i) \prod_{\{j|j \text{ visible}\}} f_j(y^i_j|\beta_j, x^i), \tag{4.2}$$

where $F_j(y^i_j|\beta_j, x^i_j)$ is the cumulative distribution function belonging to the density $f_j(y^i_j|\beta_j, x^i_j)$. Thus, conditional independence eliminates the multiple numerical integration. But it does even more: Working with a logarithmised objective function, the products in (4.2) turn into sums where each addend only depends on the parameter vector of the j-th component β_j. As a consequence, the parameter vector of the j-th component can be determined independently from the parameter vectors of other components. This allows the estimation of β_{firm} without having to determine the coefficient vectors of other expected pay supplements. So the conditional independence reduces a multidimensional to a one-dimensional problem. The objective function to identify β_j is:

$$\sum_{\{i|v_j(y^i)=1\}} \log\left(f_j(y_j^i|\beta_j, x^i)\right) + \sum_{\{i|v_j(y_i)=0\}} \log\left(F_j(y_j^i|\beta_j, x^i)\right). \qquad (4.3)$$

Formula (4.3) bears a lot of similarity with the objective function used for tobit estimation. Indeed similar principles are at work: Whenever y_j is visible it is used directly in the density, whenever it cannot be observed, it contributes by the probability of this event to occur. Different from the ordinary tobit estimator, visibility is not determined by a fixed but by a varying threshold. Thus, it is very well possible that the same value of y_j is observed for some worker i but not for another. Figure 4.3 illustrates this phenomenon and relates it to the objective function of the respective estimator. Since (4.3) respects the varying nature of the threshold, it is called *random threshold estimator*.

Suppose that expected firm performance pay is a linear function of task descriptors and control variables X and that this linear function is blurred by a normally distributed error term to yield the actual firm performance pay Y_{firm}:

$$Y_{\text{firm}} = X\beta_{\text{firm}} + \epsilon, \quad \epsilon \sim I\!N(0, \sigma^2).$$

The latent variable Y_{firm} only generates visible observations y_{firm} if it exceeds two thresholds: It must be larger than zero and larger than the fifth-largest pay supplement observation.

Due to the normality of the noise term, density and cumulative distribution function in (4.3) can be replaced by the normal density $\phi(\cdot)$ and normal c.d.f. $\Phi(\cdot)$: $f_j(t|\beta_j, x^i) = \phi(t|x^i\beta_{\text{firm}}, \sigma^2)$ and $F_j(t|\beta_j, x^i) = \Phi(t|x^i\beta_{\text{firm}}, \sigma^2)$, where the first parameter denotes the expected value of the normal distribution and the second the variance. Then, the contribution for each observation i can be computed numerically. A STATA-programme, which implements the calculation of the contributions and which is used for the following estimations, can be found in Appendix B.

4.4 Results

Matching the establishment positional data of the ECMOSS with the worker's positional data leaves about 8,000 observations. This number reduces further when variables which are used for estimation have missing values. Estimations are carried out using the random threshold

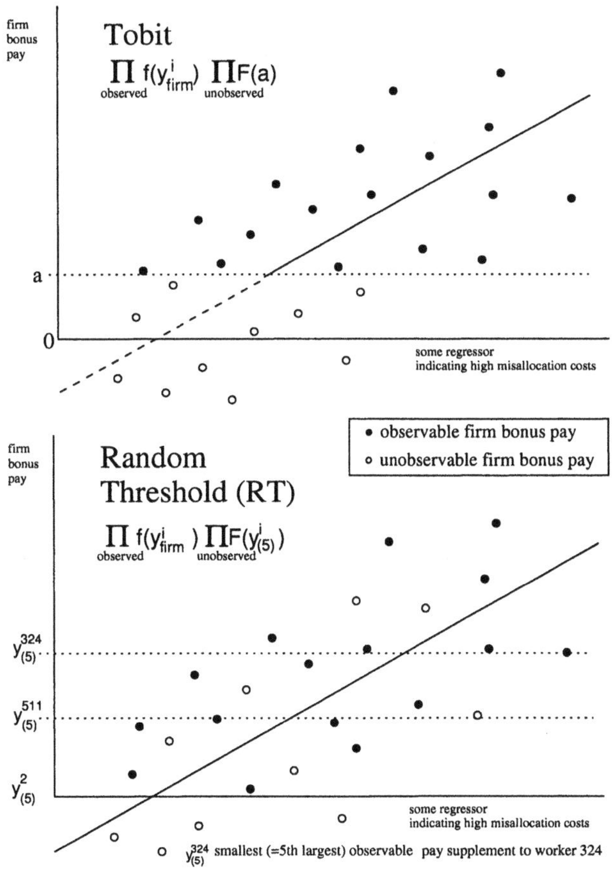

Fig. 4.3. Estimation method

In typical censoring models, values below a fixed threshold cannot be observed. Then, a consistent estimator is the tobit estimator. Here, the threshold varies, which is taken into account by the random threshold estimator.

estimator described above. Several specifications of the random threshold models are estimated with varying sets of control variables and samples. A summary of the most important results are discussed in this section and are listed in Tables 4.3 and 4.4. Tobit estimation is also carried out – primarily to generate meaningful starting values for the random threshold model. Although the coefficients of the tobit estimator slightly differ in size and significance from the random threshold model, the general findings point in the same direction (see Table C.4).

Table 4.3. Random threshold estimation with differing controls

	Intensity of firm performance pay					
	rank controls (1)		rank+risk (2)		rank+risk+firm id (3)	
	coeff.	P-value	coeff.	P-value	coeff.	P-value
SUBORD1	-0.041	0.708	-0.084	0.495	-0.004	0.966
SUBORD2	-0.130	0.305	-0.213	0.137	0.232	0.024 **
SUBORD3	-0.134	0.358	-0.176	0.285	-0.130	0.270
EXECUTE	0.074	0.283	0.020	0.804	-0.118	0.045 **
RESPONS	0.307	0.000 ***	0.336	0.000 ***	0.002	0.977
COOPERATE	0.314	0.000 ***	0.309	0.003 ***	-0.002	0.979
BASEPAY	0.118	0.000 ***	0.104	0.000 ***	0.090	0.000 ***
BASEPAY2	-0.001	0.000 ***	-0.001	0.000 ***	-0.000	0.000 ***
MALE			0.411	0.000 ***	0.026	0.707
			culture (2) education (6) age (9) dummies			
					firm dummies (245)	
Log likelihood	-4971		-4255		-1927	
Observations	6931		5860		1633	

significance levels *** 1 % ** 5 % * 10%

4.4.1 Task Descriptors

Task descriptors are variables of paramount interest in the analysis because they are viewed as proxies for misallocation costs.

Number of Subordinates

For the number of subordinates, two effects may be present. First, having many subordinates implies multiple tasks and the allocation of effort amongst them. Second, it is an indicator for the power of the respective worker on the firm performance measure. According to both effects, firm performance pay should increase in the number of subordinates. The coefficients of SUBORD1 to SUBORD3 exhibit no clear indication for a rise of firm performance pay, except that workers with five to fifteen subordinates get more firm performance pay than workers without subordinates; but this result is only weakly significant (Table 4.4, specification (3)). With respect to the interpretation that subordinates are an indicator for the costs of misallocation of effort, the finding can be explained in three ways: Either subordinates do not increase these costs beyond the degree which is captured by other task descriptors, or the supposed relationship between misallocation and firm performance pay is not valid, or this relationship is more involved. Concerning the interpretation that subordinates proxy the influence of the worker on firm performance, the weak significance might be attributed to the fact that the base payment variable already sufficiently captures this influence. As discussed later, the base payment is strongly significant and has the required sign.

Coordination and Cooperation

The professional social interaction between departments without the superior hierarchical layer being involved (COOPERATE) is only possible in large firms. At the same time, large firms are more likely to have some form of firm performance pay system because the fixed costs of establishing such a system are only covered when there is a sufficient number of workers. This can be confirmed looking at a simple two-way table where firm size categories are plotted against the existence of firm performance schemes (see Table C.3 in Appendix C). Thus, it appears as if social interaction and firm performance pay are related, where indeed this relation is most likely driven by a third variable: the size of the firm. That means, the relation between COOPERATE and firm performance pay is spurious. Once the sample is restricted to firms in which at least one worker received firm performance, the coefficient on the social interaction variable becomes insignificant (see Table 4.4, specification (4)).

Autonomous Problem-solving

A similar argument can be used for autonomous problem-solving without referring to the hierarchy (RESPONS). Consider a small firm in which the worker is under the direct supervision of the employer and in which the fixed costs of installing a firm performance pay scheme outweigh the benefits. As a consequence, we observe a positive association between having no autonomy and getting no firm performance pay. But this association has nothing to do with the hypothesis that firm performance pay is used to warrant misallocation when the worker has discretionary power. Again, firm characteristics would drive both: the use of firm performance pay and the degree of autonomy. Limiting the sample to firms where at least one worker in the sample received firm performance pay reduces the danger of this spurious relation. Indeed, the coefficient for the variable indicating autonomous problem solving drops once the sample is limited (see Table 4.4). The remaining

Table 4.4. Random threshold estimation with differing samples

Intensity of firm performance pay				
	full sample (2)		FPP>0 for some worker (4)	
	coeff.	P-value	coeff.	P-value
SUBORD1	-0.084	0.495	-0.070	0.589
SUBORD2	-0.213	0.137	0.115	0.458
SUBORD3	-0.176	0.285	-0.183	0.282
EXECUTE	0.020	0.804	-0.161	0.054 *
RESPONS	0.336	0.000 ***	0.174	0.075 *
COOPERATE	0.309	0.003 ***	-0.064	0.570
BASEPAY	0.104	0.000 ***	0.103	0.000 ***
BASEPAY2	-0.001	0.000 ***	-0.001	0.000 ***
MALE	0.411	0.000 ***	0.049	0.569
	controls for culture (2) education (6) age (9)			
Log likelihood	-4255		-2571	
Observations	5860		1633	

significance levels *** 1 % ** 5 % * 10%

correlation is also spurious: If firm dummies are introduced, the effect vanishes altogether (see Table 4.3, specification (2) and (3)).

Tightness of Task Description

When analysing the complete sample including firms in which no (sampled) worker receives performance pay, no relationship between the tightness of task description and firm performance pay can be found. Workers with general objectives get firm performance pay as often as workers who get a precise description of tasks which ought to be executed. Again, the sample composition seems to be crucial for this finding. There are several firms in which no firm performance pay occurs. In these firms, there are probably workers with general objectives and precise descriptions. This variation in the type of task is not reflected in firm performance pay; the relation cannot be observed because prohibitively large fixed costs of installing a firm performance pay scheme dominate. To support this explanation, one can limit attention to firms for which it is known that they remunerate at least one worker according to firm performance measures. This sample – which has already been mentioned previously – is the most appropriate to study the degree to which firm performance measures are utilised in relation to tasks because the firms in this sample have incurred the costs of installing a firm performance pay scheme and can now vary its application according to the tasks at different positions. In particular, they can attach higher firm performance pay to positions with global objectives (EXECUTE=0) and lower firm performance pay to positions with a precise description of tasks (EXECUTE=1). The data supports this explanation: The coefficient for the dummy which indicates a tight description of tasks is negative and significant (see Table 4.4, specification (4)).

4.4.2 Control Variables

The effects of some control variables are described in the following. They were added to the regression to eliminate other effects which also drive the relationship between firm performance pay and tasks. Although their coefficients are not the focus of attention, some interesting findings deserve to be discussed.

Base Pay

All regressions control for base pay. The base pay received by the worker is positive and always found highly significant: Workers with higher base pay get more firm performance pay. There are several explanations for this phenomenon. First, workers who get more expected firm performance pay may shoulder a higher uncertainty, which needs to be compensated by a higher uncertainty premium, that is higher base pay. Second, base salary is a proxy for rank: A higher base salary is usually associated with a higher rank in the hierarchy which in turn often implies that the influence on firm performance is higher. If the variation under control of the worker increases in comparison to the variation beyond the control of the worker, the use of the firm performance measure becomes relatively cheaper. Third, there may be an endowment effect. Suppose a worker who disposes of better financial means must be motivated by higher payments in comparison to a worker who has less financial resources. Then, the former needs to receive a higher base pay to be willing to take up the job and a higher firm performance pay to be induced to exert effort.

Whatever effects determine the positive relationship between base and firm performance pay, they are less pronounced for larger values of base pay: The coefficient for the squared base pay is negative in all regressions; this suggests a concave relationship.

Age

With respect to the age of the worker, the relation seems to follow an inverted U-shape. Young and old workers get less firm performance pay than middle-aged workers. Since we are looking at a cross section, cohort effects cannot be distinguished from age effects. Hence, older workers might get less firm performance pay either because they are more risk-averse or because their job was created and filled when firm performance pay was not yet widely spread. The latter (cohort) explanation assumes stickiness of the contract; otherwise, the form of the first contract would not matter for later contracts.

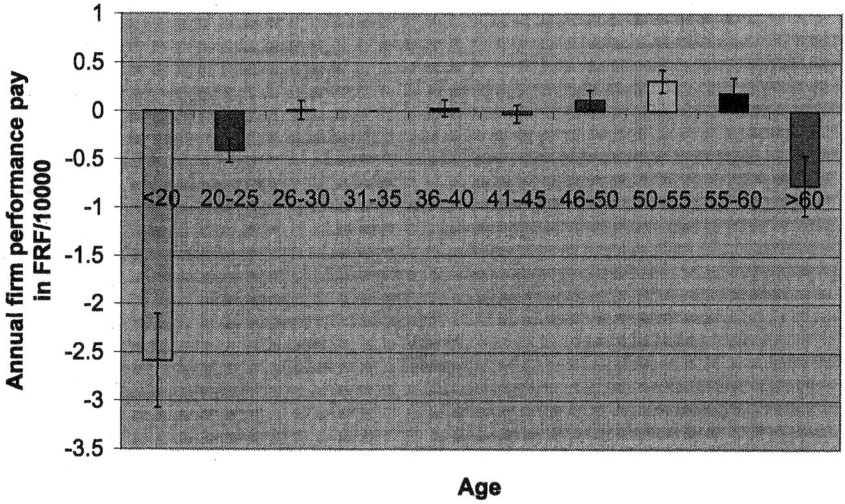

Fig. 4.4. Performance pay and age

The age-group 31-35 serves as a reference. The intervals around the bars indicate plus and minus one standard deviation.

Wage profiles rising in seniority can be explained by their incentive effects on young and middle-aged workers (Lazear 1979). But the informal incentive mechanism of continued employment and promotion fails for workers approximating the retirement age. To overcome this problem, deferred compensation could be substituted by other incentives such as explicit firm performance pay. As firm performance pay falls rather than increases, such a substitution does not seem to be present in the data (see Figure 4.4).

Young workers may possibly fill positions characterised by less discretion and less influence on firm performance pay. Since only some aspects of the position are controlled for, the low firm performance pay for young workers might be due to such positional effects.

Gender

Goldin (1986) argues that the more abbreviated and discontinuous labour market participation of female workers (due to child-bearing etc.) makes piece rate schemes more attractive to them than incentive schemes, which condition promotion or employment on past performance. In addition, piece rate schemes do not rely on subjective assessments by supervisors and thereby avoid the possibility of male supervisors biasing the evaluation in favour of their gender. Like piece rate schemes, remuneration based on firm performance measures is a more objective form of remuneration and gives immediate rewards rather than deferred compensation. So the model of Goldin (1986) can be applied to the question of whether to use firm performance measures or an implicit (work-life) incentive scheme.

Consequently, women are, *ceteris paribus*, expected to get more firm performance pay than men. Estimations on the full sample show the opposite: Male workers get significantly more firm performance pay (see Table 4.3, specification (2)). Once more, the sample under consideration plays an important role; when looking at firms which offer firm performance pay to at least one worker in the sample, the effect vanishes (see Table 4.4). This suggests that men are not *per se* getting more firm performance pay but are with a higher probability employees at firms offering firm performance pay.

Goldin (1986) finds that industries with a large proportion of female workers also have a larger fraction of piece rate workers. The discrepancy between her findings and the results from the present analysis can be traced back to several differences in the empirical strategy and data. Goldin analyses piece rates and not firm performance pay, she explains variation between industries and not between positions, and she employs data from a different country (United States vs. France) and time (1890 vs. 1992).

Firm Effects

The product manufactured or the service rendered by the firm, the market conditions, the available production technology, and the degree to which effort is observed may change considerably from firm to firm. These determinants potentially influence both, tasks as well as the costs and benefits of firm performance pay beyond the relation on the positional level. As a consequence, we observe links between tasks and firm performance pay which are not related to the theory of misallocation costs. Firm size has been given as an example to influence both, the degree of inter-departmental communication as well as the costs of setting up a firm performance pay scheme, thereby obscuring the positional relation.

To avoid effects resulting from firm size and other firm parameters, one could include the respective variables in the regression. For three reasons, this is not pursued here. First, standard errors for the coefficients of firm variables need to be corrected because they are constant for all workers from the same firm and the degree of variation in the relationship appears smaller than it is – an observation made by Moulton (1986, 1990) for linear models. Thus, one could use firm variables to control for the respective firm effects, but it is not possible to test the significance of these effects. For the correct standard errors, the correction method proposed by Moulton must be adapted from the linear model to the present non-linear situation. Second, even if observable firm variables are controlled for, there might still be unobservable heterogeneity between firms which drives the results. The third issue is a practical one: Matching the firm level data to the individual data and eliminating observations with missing values decreases the number of observations considerably.

The fact that there are several observations for most firms opens up a different avenue to control for firm effects, namely employing firm fixed-effects. Using this method, two of the above problems are avoided: Standard errors are valid and all heterogeneity between firms is accounted for. However, the nature of the censoring requires at least one observation of firm performance pay per firm to identify the firm effect. Otherwise, the firm effect can be any large negative number which prevents firm performance pay from exceeding the threshold of observability. To solve the identification problem, it is possible to work only with the sample of firms which paid at least one worker some money according to firm performance. This sub-sample has about 4,000 observations less than the original sample, but it has the additional virtue that the included firms have made the general decision to use firm performance pay.

In a next step, the estimated firm effects could be explained by firm determinants such as number of employees, industry, etc. There are three reasons why this line of analysis is not pursued in the following. First, it is better to use the actual information about the existence of firm performance schemes rather than the payment variable and this information is available on the firm level. Second, exactly this information has been already exploited by Dupuy and Lafranchi (1998) to analyse the use of performance schemes. Third, there is no obvious variable related to the costs of misallocation of effort on the firm level, so that the theory developed above has no prediction which could be tested.

4.5 Summary

If firm performance measures reduce misallocation costs, which was hypothesised in the third chapter, firm performance pay should be positively related to indicators for misallocation costs. Since misallocation concerns the behaviour of an individual worker, positional data is used to test this hypothesis.

Misallocation costs are proxied by the number of subordinates, by the question whether professional interaction with other departments is required, whether problems have to be solved autonomously, and whether the task description is precise. The first three variables point to increased misallocation costs, whereas the last indicates low misallocation costs. Indeed, interactions with other departments and autonomous problem-solving are significantly associated with higher firm performance pay, but this seems to be a spurious relation which is driven by firm characteristics such as size of the firm or the existence of a firm performance pay scheme within the firm. If a firm pays at least one of the sampled workers by firm performance, this implies that the firm has already incurred the fixed cost of setting up a firm performance pay scheme. Then, the decision about the level of expected firm performance pay at a position is not obscured by firm wide considerations. It is purely driven by the considerations for the individual position for which the hypothesis has been formulated. Restricting the sample accordingly shows that a precise description of tasks is significantly negatively associated with firm performance pay. For the other variables, the predicted effect cannot be found – probably because they proxy misallocation costs too imprecisely.

In contrast to other empirical studies, which derive predictions for very specific environments and use respective data, the data here encompass the whole private sector in France. Despite the general character of the present study and the corresponding high variability, the hypothesis that firm performance measures are used to decrease misallocation costs gets some support by the data.

5

Conclusion

The prevailing theme in hidden action models has long been the conflict between insurance and incentives. According to this theme, a completely insured worker lacks the incentive to exert costly effort, which benefits the employer and which is not contractible. On the other hand, a worker whose wage fully depends on noisy output, has to cope with uncertainty, which he or she dislikes. The desire of the employer to induce effort and the demand of the worker to be insured need to be balanced. As a result, a compromise is found and the worker is partially insured.

In this setting, all costs result from the uncertainty of output and the risk-aversion of the worker. If either of the two features is dropped, the problem disappears. Correspondingly, a performance signal has a positive value if it reduces uncertainty: The worker needs less insurance, and higher levels of effort can be induced. Thus, the worker's wage should be made a function of such a signal.

This theoretical prediction of the simple hidden action model is at odds with reality. Frequently, observable signals are not used to determine workers' wages although they are easily accessible: Truck drivers are not paid by driving distance and secretaries not by the number of keystrokes. In fact, a large fraction of labour contracts do not stipulate the use of any performance signals. This complete absence cannot entirely be explained by monitoring costs because it also concerns signals which is readily available and does not require the installation of special equipment or the hiring of a supervisor.

Apparently, the value of signals depends on more than merely a reduction in uncertainty. In particular, using signals seems to entail costs which are not captured by the single-effort hidden action model. This observation has led to the formulation of multitasking models. In this model class, employing signals leads to costs because it attracts attention away from important to less important tasks. The employer renounces utilising information in order not to distort the allocation of effort between tasks.

Multitasking models identify the cause for the diminished value of signals in reality: Using them leads to effort misallocation. This identification, however, is still vague. Several question remain unanswered: What are the reasons for the misallocation? How does the misallocation manifest itself? Finally, what are the costs of misallocation?

This monograph provides four innovations. First, the value of signals is redefined to encompass misallocation; this allows us to answer the three preceding questions. Second, this concept is applied to construct a misallocation-uncertainty trade-off model, which generates new insights into the utilisation of signals. Third, a unique data source is used to examine the predictions of the misallocation-uncertainty trade-off model empirically. Finally, in response to a censoring problem affecting this data, a class of estimators is devised which applies to most censoring problems and leads to consistent results.

The first contribution of this monograph is the definition of the costs of misallocation. In contrast to other attempts to measure misallocation, the definition is not limited to particular signal types, benefit functions, or contracts. Rather, misallocation costs can be computed for almost any hidden action model. In future work, this will allow to unifying and structuring the results of different multitasking models. Furthermore, misallocation costs can help to explain when and why signals are used or discarded.

The definition of misallocation costs attaches a number to the idea that misallocation of effort is problematic and prevents the utilisation of information. In other words, the vague and abstract idea that misallocation can be harmful is captured by misallocation costs in a simple and general form. Another conceptual advantage lies in the fact that misallocation is measured by its costs. This permits a direct comparison with other cost components such as compensation for uncertainty.

Further, two sources of misallocation costs are uncovered: uncertainty and feasibility. In the first case, misallocation is voluntarily accepted to reduce uncertainty; in the second case, it is unavoidable.

This typology of misallocation costs allows the decomposition of the loss which results from inducing effort by an incentive scheme rather than specifying effort directly; the direct specification is precluded by the non-contractibility of effort, but it provides a reference as to what could maximally be achieved. The decomposition of this loss has several merits. For instance, incentive schemes can be compared not only with respect to the loss but also with respect to the sources of the loss. For example, the loss of one incentive scheme may result from high uncertainty, while the loss of another is due to misallocations. If the value generated by an incentive scheme is identified with the value of the signals used in this incentive scheme, a similar decomposition can be undertaken for the value of signals: The net effect of including a new signal in the incentive scheme can be attributed to changes in different constituents; the value of the signal may lie in reducing the compensation for uncertainty, uncertainty-induced misallocation, feasibility-induced misallocation, or other costs. The detailed view on the value of signals leads to a better assessment of this value and may help in the future analysis of incentive problems.

Future research can fruitfully extend the concept of misallocation costs. Misallocation costs are defined in levels while many economic concepts, for example the concept leading to the information principle, are founded on marginal considerations. Thus, it would be interesting to construct marginal misallocation costs. A possible way is to start with the observation that all types of misallocation costs arise from a constraint on effort allocation. While the absolute costs are the willingness to pay for eliminating the constraint, marginal costs are the willingness to pay for a marginal relaxation of the constraint. In essence, marginal misallocation costs could be defined as the shadow price related to the constraint.

The second contribution is based on an application of the concept of misallocation costs. This application provides a new view on the relative advantage of effort versus benefit signals. In a simple model, the effort signal is taken to be the sum of efforts while the benefit signal is a noisy signal, the expected value of which coincides with the benefit of the employer. Applying the concept of misallocation costs, it is shown that the value of the effort signal lies in uncertainty reduction and the virtue of the benefit signal in misallocation reduction. The model thus exhibits an interesting trade-off between low uncertainty (if the effort signal is used) and low misallocation (if the benefit signal is used).

The proposed model does not only illustrate the trade-off between effort and benefit signals; it also explains how the balance between the two measures varies with the benefit function: The misallocation resulting from the effort measure does not always lead to the same misallocation costs. The benefit function can be relatively robust with regard to the misallocation, which means that the employer does not care which effort is exerted. However, the benefit function may also be very sensitive to misallocation. In the latter case, misallocation translates into higher misallocation costs. This tilts the balance in favour of the benefit signal.

Yet another virtue of the proposed model is that it yields an empirical prediction: If the benefit or production function is more sensitive to effort allocation, one expects to see those performance measures more intesively used which are closely related to the employer's benefit. In reality, these measures typically concern the performance of the whole firm, such as sales or profit. Measures on more disaggregated levels tend to distort effort because they do not capture all consequences of the worker's doings such as helping other workers.

Measures which aggregate the effort of different workers potentially lead to another type of misallocation: free-riding. A critical re-assessment of the respective theoretical literature shows that the free-riding effect has not been identified in multiple agent hidden action models; rather it seems to be absent from this class of models. This observation is little known and contradicts the common belief that aggregate measures generally entail free-riding. Certainly, the question whether free-riding effects occur in multiple agent hidden-action models deserves a more detailed treatment in future research.

The third achievement is the empirical analysis of the value of signals in incentive schemes. The theoretical prediction that firm performance measures have a higher value and are more intensely used when misallocation is costly is operationalised and examined using a unique data set. This study is novel in many respects: It relates task descriptors and firm performance pay on the basis of individual jobs. Previous studies either used more aggregated entities such as firms, occupational groups, or industries or were confined to a single firm. Although the data are on individual jobs, they include observations from different firms which are representative of the French private sector. Thus, results are more general than in the analysis of specific firms or industries. In addition, the analysis is not limited to the often studied upper part of the hierarchical pyramid, but includes workers on all hierarchical levels.

The generality of the data set comes at the price that no very specific hypothesis can be tested; effects have to be robust across rather different environments, and controlling for these differences is only partially possible. Nevertheless, support is found that the possibility to misallocate effort is positively associated with firm performance pay, which is taken to reflect the intensity to which firm performance measures are used.

To extend the empirical evidence on the effects of misallocation, more specific modelling is required to generate more precise implications which can then be tested using other data sets. In addition, the misallocation-uncertainty trade-off model can be used to yield predictions about other signals than firm performance measures. Finally, other models building on misallocation costs could be devised and tested.

The fourth and final contribution is the development of an estimator for censoring problems. The development was necessary to render the consistent estimation of the relationship between firm performance pay and task descriptors feasible. Thus, it is no mere theoretical construct, but serves an immediate practical purpose.

Although being motivated by a specific censoring problem, the class of censoring problems to which the estimator applies is rather general. It not only includes most censoring problems which are used in econometrics, such as the one-sided and two-sided censoring of univariate continuous variables, but goes beyond these and encompasses multivariate censoring based on almost arbitrary conditions. For some censoring problems, the estimator coincides with known estimators.

When dealing with censoring problems, it is common practice in econometrics to compose "likelihood" functions from density and probability contributions. Since the resulting objective function is no likelihood in the original sense – as it is neither a density nor a probability function – the resulting estimator is, strictly speaking, not a maximum likelihood estimator. Consequently, it is not clear that it features the respective properties, except for the one-sided censoring case for which the tobit estimator is known to be consistent, asymptotically normal and asymptotically efficient. Nonetheless, many econometricians seem to take these properties for granted. The statistical tools developed here justify this attitude to some degree: If the objective function based on density and probability contributions can be written in a particular form and if certain regularity conditions are fulfilled, then the maximiser of the objective function is identical to the derived censoring estimator, which has the usual properties of a maximum likelihood estimator.

Some of the regularity conditions for the developed estimator are not formulated in terms of the original density, but in terms of conditional densities. Future research related to the censoring estimator could aim at reformulating and simplifying these conditions.

Concluding: This monography provides new statistical and conceptual tools. While the former allow the reader to consistently estimate under many censoring problems, the latter may help to model, to better understand, and to improve the design of incentive schemes.

A

Properties of the Multivariate Censoring Estimator

This appendix proves that the estimator defined by (4.1) is consistent (Section A.1) and root-n asymptotically normal (Section A.2) and efficient (Section A.3) under mild conditions. It also explains how the estimator can be adapted to the regression context (Section A.4).

A.1 Consistency

In this section, consistency of the estimator defined by (4.1) will be proven using a standard result on the consistency of M-estimators. As $\hat{\theta}_n$ is the maximiser of any monotone transformation of the objective function in (4.1), one can alternatively work with the following objective function:

$$Q_n(\theta) := \frac{1}{n}\sum_{i=1}^{n} \log \tilde{f}_{s_i}(v^{s_i}y^i, \theta) \tag{A.1}$$

and use the machinery of M-estimation to determine its properties and in particular whether it is consistent. This can be achieved employing the following standard result (see e.g. Amemiya 1985 or Newey and McFadden 1996):

Proposition 1 (Consistency of M-estimators). *If there are measurable functions $Q_n(\theta)$ and a non-stochastic function $Q_0(\theta)$ such that (i) $Q_n(\theta)$ converges uniformly in probability to $Q_0(\theta)$, (ii) $Q_0(\theta)$ is continuous, (iii) $Q_0(\theta)$ is uniquely maximised at θ_0, and (iv) the parameter space is compact, then $\hat{\theta}_n$ is consistent for θ_0: $\hat{\theta}_n \xrightarrow{p} \theta_0$.*

The rest of this section will be devoted to finding primitive conditions for (i) to (iii) to hold.

The objective function needs to converge to the non-stochastic function $Q_0(\theta)$. Before being able to find the maximiser of the limit of the objective function, we have to ensure that this function exists and is finite. Thus, define the condition:

$$\forall s : \forall \theta : \int_{\mathcal{V}^s} \log \left\{ \tilde{f}^s(\nu^s y, \theta) \right\} \tilde{f}^s(\nu^s y, \theta_0)\, d(\nu^s y) < \infty, \qquad \text{(FIN)}$$

Given this condition, it is possible to state the following result.

Proposition 2 (Convergence). *Given condition* (FIN), $Q_n(\theta)$ *converges uniformly in probability to*

$$Q_0(\theta) = \sum_s \int_{\mathcal{V}^s} \log \left\{ \tilde{f}^s(\nu^s y, \theta) \right\} \tilde{f}^s(\nu^s y, \theta_0)\, d(\nu^s y). \qquad \text{(A.2)}$$

Proof. Since observational units are drawn independently, $Q_n(\theta)$ is the mean of independent random variables, the law of the large numbers applies, and the mean converges in probability to its expected value. The finiteness is then ensured by (FIN).

Next, continuity of the limiting objective function $Q_0(\theta)$ has to be ensured so that stochastic convergence of the argument leads to stochastic convergence of the values of the function.

Proposition 3 (Continuity). *Given that* $f(y, \theta)$ *is continuous in* θ (CON), $Q_0(\theta)$ *is continuous in* θ.

Proof. If $f(y, \theta)$ is continuous in θ, $\tilde{f}^s(\nu^s y, \theta)$ is continuous in θ, and so is $Q_0(\theta)$.

As in the case of maximum likelihood estimation, it must be possible to extract the desired information about parameters from the observations. If two different parameter values generate exactly the same observations, we can never identify which parameter value led to this observation. In other words, it is required that parameters can be identified from observations. Define the statistical model to be *identified* if and only if

$$\forall \theta \neq \theta' \; \exists s : \; P(S = s) > 0 : \; f^s(\nu^s y, \theta) \neq f^s(\nu^s y, \theta'). \qquad \text{(ID)}$$

In other words, at least one state must exist under which differences in the parameter translate into differences in the conditional density. Similar, to the identification condition in maximum likelihood estimation, this condition may be difficult to verify. The next result proves that the parameter is indeed uniquely determined when the condition can be verified.

Proposition 4 (Unique maximiser). *Under* (ID) *and* (FIN), $Q_0(\cdot)$ *is uniquely maximised at the true parameter* θ_0.

The proof for this result is very similar to the proof of the uniqueness of the maximiser of the limiting objective function when working with likelihoods.

Proof. Consider the difference between the limiting objective function evaluated at the true parameter, $Q_0(\theta_0)$, and at a different parameter $Q_0(\theta)$:

$$\begin{aligned} Q_0(\theta_0) - Q_0(\theta) &= \mathrm{E}_Y\left(\log\left\{\tilde{f}^s(\nu^s y, \theta_0)\right\} - \log\left\{\tilde{f}^s(\nu^s y, \theta)\right\}\right) \\ &= \mathrm{E}_Y\left(-\log\left\{\frac{\tilde{f}^s(\nu^s y, \theta)}{\tilde{f}^s(\nu^s y, \theta_0)}\right\}\right) \\ &> -\log\left\{\mathrm{E}_Y\left(\frac{\tilde{f}^s(\nu^s y, \theta)}{\tilde{f}^s(\nu^s y, \theta_0)}\right)\right\}, \end{aligned} \tag{A.3}$$

where the last inequality follows from the strict version of Jensen's inequality for non-constant random variables. By (ID), the expected value is indeed taken over a non-constant random variable. As

$$\mathrm{E}_{Y|S}\left(\frac{\tilde{f}^s(\nu^s y, \theta)}{\tilde{f}^s(\nu^s y, \theta_0)},\right) = 1$$

we get $Q_0(\theta_0) - Q_0(\theta) > 0$, and θ_0 is the unique maximum.

The consistency of the estimator (4.1) can now be deducted from the conditions (FIN), (CON), and (ID) which together with the Propositions 2 to 4 imply that the requirements of the standard consistency theorem are valid.

Theorem 1 (Consistency of the estimator). *If there are measurable functions* $Q_n(\theta)$, (FIN), (CON), (ID) *hold, and the parameter space is compact, then* $\hat{\theta}_n \xrightarrow{p} \theta_0$.

A.2 Asymptotic Normality

If one takes root-n of a maximum likelihood estimator, this estimator is asymptotically normally distributed and asymptotically efficient under regularity conditions. In this section, conditions are derived under which (4.1) has these properties. Again, the conditions resemble the respective conditions for maximum likelihood estimators. This is no coincidence because the proof is based on a standard result for M-estimators (Theorem 4.1.3 in Amemiya (1985), where assumption B is replaced using Theorem 4.1.5):

Proposition 5 (Asymptotic normality of M-estimators). *If the following conditions hold: (i) $\hat{\theta}_n$, the maximiser of $Q_n(\cdot)$, is consistent for θ_0, (ii) θ_0 lies in the interior of the parameter space Θ, (iii) Q_n is twice continuously differentiable in an open and convex neighbourhood $\mathcal{N}$ of θ_0, (iv) $\sqrt{n}\nabla_\theta Q_n(\theta)|_{\theta=\theta_0} \xrightarrow{d} N(0,J)$, (v) $\nabla_{\theta\theta} Q_n(\theta)|_{\theta=\hat{\theta}_n} \xrightarrow{P} H(\theta_0)$ with $H(\theta)$ finite, non-singular, and continuous at θ_0, it follows that*

$$\sqrt{n}(\hat{\theta}_n - \theta_0) \xrightarrow{d} N(0, H^{-1}JH^{-1}).$$

Under the assumptions of Theorem 1, condition (i) is valid. Condition (ii) ensures that the maximum is not a corner solution and hence that the first derivative of $Q_0(\cdot)$ disappears at θ_0. Subsequently, conditions (iii) to (v) should be replaced by primitive conditions on the density $f(y,\theta)$.

Begin by assuming:

$$f(y,\theta) \text{ is twice continuously differentiable at } \theta_0. \qquad \text{(DIFF)}$$

Additionally, the differentiation operators should be exchangeable with integration, which is, for example, fulfilled if the support over which is integrated does not depend on θ:

$$\nabla_\theta \int f(y,\theta)dy = \int \nabla_\theta f(y,\theta)dy \quad \nabla_{\theta\theta} \int f(y,\theta)dy = \int \nabla_{\theta\theta} f(y,\theta)dy, \qquad \text{(EID)}$$

where the equalities are required to hold only evaluated at $\theta = \theta_0$. The exchangeability can be used to compute the first and second derivative of $\tilde{f}^s(\nu^s y, \theta)$ with respect to θ at θ_0:

$$\begin{aligned} \nabla_\theta \tilde{f}^s(\nu^s y,\theta)\Big|_{\theta=\theta_0} &= \int_{\mathcal{V}^s} \nabla_\theta \tilde{f}(y,\theta) d(\bar{\nu}^s y)\Big|_{\theta=\theta_0}, \\ \nabla_{\theta\theta} \tilde{f}^s(\nu^s y,\theta) d(\bar{\nu}^s y)\Big|_{\theta=\theta_0} &= \int_{\mathcal{V}^s} \nabla_{\theta\theta} \tilde{f}(y,\theta) d(\bar{\nu}^s y)\Big|_{\theta=\theta_0} \end{aligned} \qquad \text{(A.4)}$$

Next, define:

$$J(\theta) := \mathrm{E}_Y \left[\frac{\nabla_\theta \tilde{f}^s(\nu^s Y,\theta) \nabla_\theta \tilde{f}^s(\nu^s Y,\theta)'}{\tilde{f}^s(\nu^s Y,\theta) \tilde{f}^s(\nu^s Y,\theta)} \Big|_{\theta=\theta_0} \right], \qquad \text{(A.5)}$$

where expectations are taken with respect to the true parameter θ_0. Later we will prove that $J(\theta)$ is the second derivative of the limit function $Q_0(\theta)$ and thus deserves to be called the Jacobian matrix. Since the second derivative must exist and must be finite at its maximiser θ_0, it is assumed that

$$J := J(\theta_0) \text{ exists and is finite.} \tag{EX}$$

Under the defined conditions, it is now possible to calculate the distribution of the first derivative of the objective function:

Proposition 6 (Asymptotic normality of the first derivative). *Under* (DIFF), (FIN), (EX), (EID), *and if* J *defined in* (EX) *is nonsingular, then*

$$\sqrt{n}\,\nabla_\theta Q_n(\theta)|_{\theta=\theta_0} \xrightarrow{d} N(0, J).$$

Proof. $\sqrt{n}\,\nabla_\theta Q_n(\theta)|_{\theta=\theta_0}$ can be rewritten as the sum of i.i.d. random variables:

$$\sqrt{n}\,\nabla_\theta Q_n(\theta)|_{\theta=\theta_0} = \frac{1}{\sqrt{n}} \sum_{i=1}^{n} \underbrace{\frac{\nabla_\theta \tilde{f}^s(v^{s_i},\theta)|_{\theta=\theta_0}}{\tilde{f}^s(v^{s_i},\theta)|_{\theta=\theta_0}}}_{=:Q^i_\nabla}, \tag{A.6}$$

and by the central-limit theorem its distribution converges to a normal distribution with mean $\mathrm{E}(Q^i_\nabla)$ and variance-covariance matrix $\mathrm{VAR}(Q^i_\nabla)$. The existence of $\mathrm{E}(Q^i_\nabla)$ is ensured by (EX) and Jensen's inequality, its value is:

$$\begin{aligned}
\mathrm{E}(Q^i_\nabla) = {} & \sum_s P(S=s,\theta_0) \int \frac{\nabla_\theta \tilde{f}^s(v^{s_i},\theta)|_{\theta=\theta_0}}{\tilde{f}^s(v^{s_i},\theta_0)} \cdot \frac{\tilde{f}^s(v^{s_i},\theta_0)}{P(S=s,\theta_0)} d(\nu^s y) \\
\overset{\text{(EID)}}{=} {} & \sum_s P(S=s,\theta) \cdot \nabla_\theta \underbrace{\int \frac{\tilde{f}^s(v^{s_i},\theta)}{P(S=s,\theta)} d(\nu^s y)}_{=1} \Bigg|_{\theta=\theta_0} = 0.
\end{aligned} \tag{A.7}$$

As the expected value is the zero vector, $\mathrm{VAR}(Q^i_\nabla) = \mathrm{E}(Q^i_\nabla (Q^i_\nabla)')$. By plugging in Q^i_∇ one immediately gets $\mathrm{VAR}(Q^i_\nabla) = J$, the existence of which is ensured by (EX).

But the assumptions (DIFF), (EX), and (EID) do not only allow to compute the limit of the first derivative when it is evaluated at the maximiser but also yield the limit of the second derivative:

Proposition 7 (Convergence of the second derivative). *Given* (DIFF), (EX), (EID), *and* $\hat{\theta}_n \to \theta_0$, *it follows that* $\nabla_{\theta\theta} Q_n(\theta)|_{\theta=\hat{\theta}_n} \xrightarrow{P} -J$, *where* J *is positive definite.*

Proof. The second derivative of the objective function with respect to θ is:

$$\nabla_{\theta\theta} Q_n(\theta) = \frac{1}{n} \sum_{i=1}^{n} \underbrace{\frac{\tilde{f}^s(v^{s_i} y_i, \theta) \nabla_{\theta\theta} \tilde{f}^s(v^{s_i} y_i, \theta) - \nabla_\theta \tilde{f}^s(v^{s_i} y_i, \theta) \nabla_\theta \tilde{f}^s(v^{s_i} y_i, \theta)'}{\tilde{f}^s(v^{s_i} y_i, \theta)^2}}_{=:Q^i_{\nabla}} . \tag{A.8}$$

By the law of large numbers, Q^i_{∇} approaches its expected value:

$$\begin{aligned} \mathrm{E}(Q^i_{\nabla}(\theta)) =& \mathrm{E}\left(\frac{\nabla_{\theta\theta} \tilde{f}^s(v^{s_i} y_i, \theta)}{\tilde{f}^s(v^{s_i} y_i, \theta)} \right) - \mathrm{E}\left(\frac{\nabla_\theta \tilde{f}^s(v^{s_i} y_i, \theta) \nabla_\theta \tilde{f}^s(v^{s_i} y_i, \theta)'}{\tilde{f}^s(v^{s_i} y_i, \theta)^2} \right) \\ =& \sum_s P(S = s, \theta_0) \int \nabla_{\theta\theta} \tilde{f}^s(v^{s_i} y_i, \theta) d(v^{s_i} y) \cdot \frac{1}{P(S = s, \theta_0)} \\ & - J(\theta) \\ \stackrel{\text{(EID)}}{=}& \sum_s P(S = s, \theta_0) \underbrace{\nabla_{\theta\theta} \int f(v^{s_i} y_i, \theta) d(v^{s_i} y)}_{=0} - J(\theta) = -J(\theta). \end{aligned} \tag{A.9}$$

As this expected value is continuous in θ around θ_0, we can use Theorem 4.1.5 in Amemiya (1985) to conclude from $\hat{\theta}_n \xrightarrow{p} \theta_0$ that $\mathrm{E}[Q^i_{\nabla}(\hat{\theta}_n)] \xrightarrow{p} \mathrm{E}\left[Q^i_{\nabla}(\theta_0)\right]$. Overall, we get $\nabla_{\theta\theta} Q_n(\theta)|_{\theta=\hat{\theta}_n} \xrightarrow{p} \mathrm{E}\left[Q^i_{\nabla}(\theta_0)\right] = -J$. As $-J$ is the second derivative of the objective function evaluated at a unique and interior maximum, it must be negative definite. So, J must be positive definite.

Using Theorem 1 and Results 5 to 7, yields:

Theorem 2 (Asymptotic normality of the estimator). *If* (ID), (EX), (FIN), (EID), *and* (DIFF) *hold, and* θ_0 *is in the interior of the compact parameter space* Θ, *then* $\sqrt{n}(\hat{\theta}_n - \theta_0) \xrightarrow{d} N(0, J^{-1})$.

A.3 Asymptotic Efficiency

A by-product of Theorem 2 is the asymptotic variance-covariance matrix of the root-n-estimator. By comparing this matrix to the Cramer-Rao lower bound, one can check for asymptotic efficiency. First, the Cramer-Rao lower bound is established to be J^{-1} :

Proposition 8. *Given* (EID) *and non-singularity of* $J(\theta)$ *as defined in Equation* (A.5), *the Cramer-Rao lower bound is* $J(\theta)^{-1}$. *The variance-covariance matrix* $VAR(T)$ *of any unbiased estimator* T *based on* $(s, \nu^s y)$ *for all* s *is smaller or equal according to the Löwner norm:*

$$\forall a: \qquad a'(VAR(T) - J(\theta)^{-1})a \geq 0. \tag{A.10}$$

Proof. Let T be an unbiased estimator for $\tau(\theta)$, a differentiable function of the parameter to be estimated. Then,

$$\nabla_\theta \tau(\theta) = \nabla_\theta \mathrm{E}(T) = \nabla_\theta \sum_s P(S = s) \int T(y) f_s(v^s y, \theta) d(v^s y).$$

Subtracting $\tau(\theta) \cdot \nabla_\theta \sum_s P(S = s) \int f_s(v^s y, \theta) d(v^s y) = 0$ and rewriting the formula using that $\tilde{f}_s(\nu^s y, \theta) = P(S = s) \cdot f_s(\nu^s y, \theta)$ yields:

$$\nabla_\theta \tau(\theta) = \nabla_\theta \sum_s \int T(y) \tilde{f}_s(v^s y, \theta) d(v^s y) - \tau(\theta) \nabla_\theta \sum_s \int \tilde{f}_s(v^s y, \theta) d(v^s y).$$

Using (EID) and writing the different terms under a common integral leads to:

$$\nabla_\theta \tau(\theta) = \sum_s \int (T(y) - \tau(\theta)) \nabla_\theta \tilde{f}_s(v^s y, \theta) d(v^s y).$$

Multiplying with $\frac{\tilde{f}_s(v^s y, \theta)}{\tilde{f}_s(v^s y, \theta)}$ in the integral results in:

$$\begin{aligned}
\nabla_\theta \tau(\theta) &= \sum_s \int (T(y) - \tau(\theta)) \frac{\nabla_\theta \tilde{f}_s(v^s y, \theta)}{\tilde{f}_s(v^s y, \theta)} \tilde{f}_s(v^s y, \theta) d(v^s y) \\
&= \mathrm{E}\left[\underbrace{(T(y) - \tau(\theta))}_{=:U} \underbrace{\frac{\nabla_\theta \tilde{f}_s(v^s y, \theta)}{\tilde{f}_s(v^s y, \theta)}}_{=:V} \right].
\end{aligned} \tag{A.11}$$

According to the Cauchy-Schwarz inequality:

$$\forall a : \quad a' \left(\mathrm{E}(UU')\mathrm{E}(VV') - \mathrm{E}(UV)\mathrm{E}(UV)' \right) a \geq 0.$$

Next, identify the different terms in this inequality with known quantities:

$$\begin{aligned}
\mathrm{E}(UU') &= \mathrm{E}\left[(T(y) - \tau(\theta)) (T(y) - \tau(\theta))' \right] = \mathrm{VAR}(T) \\
\mathrm{E}(VV') &\overset{(A.5)}{=} J(\theta) \\
\mathrm{E}(UV) &\overset{(A.11)}{=} \nabla_\theta \tau(\theta).
\end{aligned}$$

If $\tau(\cdot)$ is the identity, then $\nabla_\theta \tau(\theta) \nabla_\theta \tau(\theta)' = I$. If in addition $J(\theta)$ is non-singular, one gets the following inequality:

$$\forall a : a' (\mathrm{VAR}(T) - J(\theta)^{-1}) \, J(\theta) a \geq 0.$$

This inequality means that the product is non-negative definite, because J is positive definite, $\mathrm{VAR}(T) - J(\theta)^{-1}$ must be non-negative definite and we obtain (A.10).

Now, the variance-covariance matrix of the root-n estimator can be compared with the Cramer-Rao lower bound. Since the variance-covariance matrix attains the bound, it can be concluded:

Corollary 1. *The estimator* $\sqrt{n}\hat{\theta}_n$ *is asymptotically efficient.*

A.4 Multivariate Censored Regression

In many applications, observational units will differ by observable characteristics X^i, which have an effect on the distribution of Y. To allow for this in the modelling framework, the formerly fixed θ is supposed to be an individual parameter which results from the interplay of observable characteristics X^i with a fixed parameter β: $\theta^i = g(\beta, X^i)$.

Since observational units are drawn randomly, the observable characteristics X^i can be modelled by a random variable. Assume this random vector to be continuously distributed,[1] so that the joint density of Y and X can be decomposed: $f_{Y,X}(y, g(\beta, x)) = f_{Y|X}(y|g(\beta, x)) \cdot f_X(x)$. Accordingly, the contribution of a particular state s becomes:

$$\tilde{f}^s(\nu^s y, \beta, x) = \underbrace{\int_{\mathcal{V}^s} f(y|\beta, x) d(\bar{v}_s y)}_{=: \tilde{f}(\nu^s y|\beta, x)} \cdot f_X(x),$$

[1] Extension to the discrete case is straightforward.

and thus the logarithmised objective function is:

$$Q_n(\theta) = \sum_{i=1}^{n} \log\left(\tilde{f}^s(v^{s_i} y^i | \beta, x^i)\right) + \sum_{i=1}^{n} \log\left(f_X(x^i)\right). \tag{A.12}$$

As the last term does not change in β, it can be ignored when maximising. So the resulting objective function closely resembles the objective function from formula (A.1), which was analysed in the preceding sections. All conditions, proofs, and theorems can be adapted to this new objective function by replacing $f(y, \theta)$ by $f(y|\beta, x)$ and $\tilde{f}^s(v^s y, \theta)$ by $\tilde{f}^s(v^s y|\beta, x)$ and requiring the respective statement to hold for all x. Additionally, one needs $\int \log(f_X(x)) f_X(x) dx < \infty$, to ensure finiteness of the limiting objective function. The identifiability condition becomes:

$$\forall \beta \neq \beta' \; \exists s, \mathcal{X} : \; P_X(x \in \mathcal{X}) > 0 : \; \tilde{f}^s(\nu^s y, \beta, x) \neq \tilde{f}^s(\nu^s y, \beta', x). \tag{ID'}$$

For the linear case $g(X, \beta) = X\beta$ and given (ID), this is fulfilled if X has full rank with positive probability.

The introduction of X has no effect on the asymptotic normality result: All proofs are based on derivatives of the objective function with respect to the parameter. Since the second sum in the objective function is independent of the parameter β, it cancels when taking derivatives.

B

Stata Programme

The following program (mcrlis) calculates the contributions for a vector of observations to the objective function which is then maximised to find parameter estimates.

```
 1: program define mcrlis
 2:     version 7.0
 3:     args lnf theta1 s
 4:     tempvar small sigma
 5:     quietly gen double ‘sigma’=exp(‘s’)
 6:     quietly gen double ‘small’=minpay
 7:     quietly replace
                ‘lnf’=norm((‘small’-‘theta1’)/‘sigma’)
                if $ML_y1<=0
 8:     quietly replace
        ‘lnf’=normden(($ML_y1-‘theta1’)/‘sigma’)/‘sigma’
                if $ML_y1>0
 9:     quietly replace
                ‘lnf’= 0.0000000000000000000001
                if ‘lnf’ < 0.0000000000000000000001
10:     quietly replace ‘lnf’=log(‘lnf’)
11: end
```

It is defined for STATA version 7.0 (line 1) and has three arguments (line 2): **lnf**, which in the end contains the contribution vector, **theta1**, which contains the linear combination of parameter and data, and **s**, which is the natural logarithm of the standard error. The temporal variables **small** and **sigma** are created to take up the smallest observable bonus payment (minpay) and the standard error (line 4). The standard error is computed from **s**, and the smallest visible bonus payment is set (line 5 and 6). Given that the dependent variable (firm performance pay,

`$ML_y1`) is not observable, the contribution, `lnf`, is set to the c.d.f. of a standard normal distribution adapted to the parameters and evaluated at the smallest visible observation (line 7). Given that the dependent variable (firm performance pay, `$ML_y1` is observable, the contribution, `lnf`, is set to the density of a standard normal distribution adapted to the parameters and evaluated at the actual observation, `$ML_y1` (line 8). To avoid undefined objects when taking the logarithm in line 10, very small values are set to a small value, which can still be evaluated by the inbuilt routine to compute the logarithm (line 9).

For the actual computation, the program mcrlis is called by the Stata procedure `ml`, which is designed to determine M-estimators for a given function which computes contributions.

C

Tables

Table C.1. Questions and variable codings

Question	Coding
Aviez-vous alors des fonctions d'encadrement? *[Do you occupy a management position?]*	1=yes, 0=no
Combien de salarié aviez-vous sous votre responsabilité? *[For how many employees are you responsible?]*	SUBORDx=0, except
1-5	SUBORD1=1
6-15	SUBORD2=1
> 15	SUBORD3=1
Comment était défini le travail á accomplir? *[How is the work that had to be accomplished defined?]*	
plutôt par une description de tâches précises à exécuter... *[mainly by a description of precise tasks to be executed...]*	EXECUTE=1
plutôt par une fixation d'objectifs globaux *[mainly by fixing general objectives]*	EXECUTE=0
En cas d'incident mineur dans la production (ou la marche du service) quelle réaction attendait-on de vous ... *[In case of a minor accident in the production (or on the market for your service) what reaction is expected from you ...]*	
...vous étiez encouragé à régler d'abord vous-même le problème *[...are you encouraged to solve the problem yourself?]*	RESPONS=1
...vous deviez en référer avant tout à la hiérarchie? *[...do you have to consult your supervisor first?]*	RESPONS=0
Aviez-vous des contact professionals avec des salariés d'autres services, sans passer par votre supérieur hiérarchique immédiat? *[Did you have professional contacts with workers from other departments without the hierarchical tier above you being involved?]*	
souvent, parfois *[often, sometimes]*	COOPERATE=1
jamais, il n'y avait pas d'autres services *[never, there were no other services]*	COOPERATE=0

Table C.2. Sample statistics

Description	Variable	Mean	Variance	Minimum	Maximum
	Annual base payments in FRF/100²				
	base-pay	10.65669	7.968272	.0103	141.3061
	Observed annual bonus payments in FRF/100² linked to				
base pay	fixbon	0.8357765	1.1470110	0	170.344
personal characteristics	nonbon	0.3128859	0.6484105	0	7.8228
constraints	conbon	0.2509512	0.9072436	0	41.2551
firm performance	peebon	0.1930310	0.8530477	0	25.044
team performance	petbon	0.0456280	0.5681872	0	29.399
individual performance	peibon	0.1336868	0.7452481	0	24.4916
special events	evebon	0.0028748	0.0799834	0	4.5
legal circumstances	sevbon	0.0429668	0.3563686	0	10.0299
	Work descriptors				
	manager	0.2316865	0.4219458	0	1
see	subord1	0.0996593	0.2995707	0	1
Table C.1	subord2	0.0715503	0.2577637	0	1
	subord3	0.0512777	0.2205823	0	1
	execute	0.5974446	0.4904544	0	1
	autonom	0.6948893	0.4604935	0	1
	social	0.7853492	0.4106148	0	1
	education				
no certificate	noed	0.2981261	0.4574741	0	1
Certificat d'etudes primaire	prim	0.2160136	0.4115587	0	1
Brevet d'etudes du 1er cycle	bepc	0.2499148	0.4330004	0	1
Baccalauréat	bacc	0.1555366	0.3624464	0	1
Diplom 1er cycle	uni1	0.0151618	0.1222068	0	1
Diplom 2eme cycle	uni2	0.0299830	0.1705548	0	1
Diplom 3eme cycle	uni3	0.0352641	0.1844622	0	1
	Gender				
	male	0.5798978	0.4936171	0	1
	non-french nationality				
other European	europe	0.0250426	0.1562678	0	1
other non-European	africar	0.0206133	0.1420979	0	1
	Age				
<20	aged1	0.0303237	0.1714910	0	1
20-25	aged2	0.0984668	0.2979701	0	1
26-30	aged3	0.1488927	0.3560130	0	1
31-35	aged4	0.1643952	0.3706654	0	1
36-40	aged5	0.1649063	0.3711276	0	1
41-45	aged6	0.1599659	0.3666057	0	1
46-50	aged7	0.1156729	0.3198595	0	1
50-55	aged8	0.0734242	0.2608537	0	1
55-60	aged9	0.0374787	0.1899479	0	1
>60	aged10	0.0064736	0.0802046	0	1

number of observations: 5870

Table C.3. Firm performance schemes and firm size

	Number of workers					
Scheme exists	1-10	11-20	21-50	51-100	101-500	>500
no	1816	1704	2746	1097	601	251
yes	523	863	1687	829	654	396

The larger the firm, the more likely the existence of a firm performance scheme. A respective χ^2*-test (5 degrees of freedom) is highly significant.*

Table C.4. Random threshold and tobit estimation

Intensity of firm performance pay				
	Tobit (3)		Random Threshold (1)	
	coeff.	P-value	coeff.	P-value
SUBORD1	-0.053	0.648	-0.041	0.708
SUBORD2	-0.184	0.177	-0.130	0.305
SUBORD3	-0.102	0.513	-0.134	0.358
EXECUTE	0.120	0.109	0.074	0.283
RESPONS	0.334	0.000 ***	0.307	0.000 ***
COOPERATE	0.333	0.001 ***	0.314	0.000 ***
BASEPAY	0.127	0.000 ***	0.118	0.000 ***
BASEPAY2	-0.001	0.000 ***	-0.001	0.000 ***
CONSTANT	-3.609	0.000 ***	-3.276	0.000 ***
Log likelihood	-5037		-4972	
Observations	6931		6931	

significance levels *** 1 % ** 5 % * 10%

References

AGHION, P., AND J. TIROLE (1997): "Formal and Real Authority in Organizations," *Journal of Political Economy*, 105(1), 1–29.

AKERLOF, G. (1970): "The Market for Lemons: Quality Uncertainty and the Market Mechanism," *Quarterly Journal of Economics*, 89, 488–500.

AKERLOF, G. A., AND L. F. KATZ (1989): "Workers' Trust Funds and the Logic of Wage Profiles," *Quarterly Journal of Economics*, 104(3), 525–536.

ALTONJI, J. G., AND R. BLANK (1999): "Race and Gender in the Labor Market," in *Handbook of Labor Economics*, ed. by O. Ashenfelter, and D. Card, vol. 3, pp. 3143–3259. Elsevier B.V., Amsterdam.

AMEMIYA, T. (1973): "Regression Analysis when the Dependent Variable is Truncated Normal," *Econometrica*, 41(6), 997–1016.

——— (1984): "Tobit Models: A Survey," *Journal of Econometrics*, 24, 3–61.

——— (1985): *Advanced Econometrics*. Basil Blackwell, Oxford.

BAKER, G. (2002): "Distortion and Risk in Optimal Incentive Contracts," *Journal of Human Resources*, 37(4), 728–751, Special Issue on Designing Incentives to Promote Human Capital.

BAKER, G., R. GIBBONS, AND K. J. MURPHY (2002): "Relational Contracts and the Theory of the Firm," *Quarterly Journal of Economics*, 117(1), 39–84.

BAKER, G., M. GIBBS, AND B. HOLMSTRÖM (1994): "The Wage Policy of a Firm," *Quarterly Journal of Economics*, pp. 921–955.

BECKER, G. (1957): *The Economics of Discrimination*. University of Chicago Press, Chicago.

BROWN, M., A. FALK, AND E. FEHR (2002): "Contractual Incompleteness and the Nature of Market Interactions," presented at IZA 23rd of April 2002.

——— (Forthcoming): "Contractual Incompleteness and the Nature of Market Interactions," *Econometrica*, presented at IZA 23rd of April 2002.

BULL, C. (1987): "The Existence of Self-Enforcing Implicit Contracts," *Quarterly Journal of Economics*, 102, 147–159.

CARMICHAEL, L. H. (1985): "Can Unemployment Be Involuntary? Comment," *American Economic Review*, 75(5), 1213–1214.

CHE, Y.-K., AND D. B. HAUSCH (1999): "Cooperative Investments and the Value of Contracting," *American Economic Review*, 89(1), 125–147.

CHE, Y.-K., AND S.-W. YOO (2001): "Optimal Incentives for Teams," *American Economic Review*, 91(3), 525–541.

DAVIDSON, R., AND J. G. MACKINNON (1993): *Estimation and Inference in Econometrics.* Oxford University Press, Oxford.

DIXIT, A., AND R. S. PINDYCK (1994): *Investment under Uncertainty.* Princeton University Press, Princeton.

DUPUY, K., AND J. LAFRANCHI (1998): "Performance absolue ou relative? Comment les etablissement francais choisisent-ils leurs primes salariales?," Discussion paper, Document de Travail ERMES 98.06.

ENCINOSA III, W. E., M. GAYNOR, AND J. B. REBITZER (1997): "The Sociology of Groups and Economics of Incentives Theory and Evidence on Compensation Schemes," Discussion Paper 5953, NBER.

FEHR, E., AND A. FALK (1999): "Wage Rigidity in a Competitive Incomplete Contract Market," *Journal of Political Economy*, 107(1), 106–134.

FEHR, E., AND K. M. SCHMIDT (2000): "Fairness, Incentives, and Contractual Choices," *European Economic Review*, 44, 1057–1068.

FELTHAM, G. A., AND J. XIE (1994): "Performance Measure Congruity and Diversity in Multi-Task Principal/Agent Relations," *The Accounting Review*, 69(3), 429–453.

FERRALL, C., AND B. SHEARER (1999): "Incentives and Transactions Costs Within the Firm: Estimating an Agency Model Using Payroll Records," *Review of Economic Studies*, 66, 309–338.

FITZENBERGER, B. (1997): "A Guide to Censored Quantile Regression," in *Handbook of Statistics, Vol 15: Robust Inference*, ed. by G. Maddala, and C. Rao, pp. 405–435. Elsevier, Amsterdam, Reprint.

FUDENBERG, D., AND E. MASKIN (1986): "The Folk Theorem in Repeated Games with Discounting or with Incomplete Information," *Econometrica*, 54(3), 533–554.

FUDENBERG, D., AND J. TIROLE (1990): "Moral Hazard and Renegotiation in Agency Contracts," *Econometrica*, 58, 1279–1319.

——— (1991): *Game Theory.* The MIT Press, Cambridge, Massachusetts.

GIBBONS, R. (1996): "Incentive and Careers in Organisations," in *Advances in Economics and Econometrics: Theory and Applications*, ed. by D. Kreps, and K. Wallis, chap. 1, pp. 1–37. Cambridge University Press, Cambridge.

GIBBONS, R., AND M. WALDMAN (1999): "Careers in Organizations: Theory and Evidence," in *Handbook of Labor Economics*, ed. by O. Ashenfelter, and D. Card, vol. 3, pp. 2373–2429. Elsevier Science B.V., Amsterdam.

GIBBS, M., K. MERCHANT, W. V. DER STEDE, AND M. VARGUS (2002): "The Structure of Incentive Contracts: Evidence from Auto Dealerships," handout at the CEPR/SITE workshop "Incentives and Beyond", Stockholm May 2002.

GOLDIN, C. (1986): "Monitoring Costs and Occupational Segregation by Sex," *Journal of Labor Economics*, 4(1), 1–27.

GOURIÉROUX, C., AND J. JASIAK (2001): *Financial Econometrics: Problems, Models, Methods.* Princeton University Press, Princeton.

GOURIÉROUX, C., AND A. MONFORT (1996): *Simulation-Based Econometric Methods*chap. 3.1, pp. 41–49. Oxford University Press, Oxford and New York.

GOURIÉROUX, C., A. MONFORT, AND A. TROGNON (1984): "Pseudo Maximum Likelihood Methods: Theory," *Econometrica*, 52, 681–700.

GREEN, J. R., AND N. L. STOKEY (1983): "A Comparison of Tournaments and Contracts," *Journal of Political Economy*, 91(3), 349–364.

GROSSMAN, S. J., AND O. D. HART (1983): "An Analysis of the Principal-Agent Problem," *Econometrica*, 51(1), 7–46.

——— (1986): "The Costs and Benefits of Ownership: A Theory of Vertical and Lateral Integration," *Journal of Political Economy*, 94, 691–719.

HART, O., AND J. MOORE (1990): "Property Rights and the Nature of the Firm," *Journal of Political Economy*, 98, 1119–1158.

HARTOG, J., AND N. JONKER (1999): "On a Simple Measure of Risk Aversion," distributed at IZA seminar, June 1999.

HECKMAN, J. J., AND R. ROBB (1985): "Alternative Methods for Evaluating the Impact of Interventions," in *Longitudinal Analysis of Labor Market Data*, ed. by J. J. Heckman, and B. Singer, chap. 4, pp. 156–245. Cambridge University Press, Cambridge.

HIRAO, Y. (1993): "Task Assignment and Agency Structures," *Journal of Economics and Management Strategy*, 2(2), 299–323.

HIRSHLEIFER, J., AND J. RILEY (1992): *The Analytics of Uncertainty and Information.* Cambridge University Press, Cambridge.

HOLMSTRÖM, B. (1979): "Moral Hazard and Observability," *Bell Journal of Economics*, 10(1).

——— (1982): "Moral Hazard in Teams," *Bell Journal of Economics*, 13, 324–340.

HOLMSTRÖM, B., AND P. MILGROM (1987): "Aggregation and Linearity in the Provision of Intertemporal Incentives," *Econometrica*, 55(2), 303–328.

——— (1991): "Multitask Principal-Agent-Analysis: Incentive Contracts, Asset Ownership, and Job Design," *Journal of Law, Economics, and Organization*, 7, 24–52.

HUCK, S., D. KÜBLER, AND J. WEIBULL (2001): "Social Norms and Optimal Incentives in Firms," Research Paper in Economics (RePEc): ≪http://swopec.hhs.se/hastef/papers/hastef0466.pdf≫.

ITOH, H. (1991): "Job Design, Delegation and Cooperation: A Principal-Agent Analysis," *European Economic Review*, 38(3–4), 691–700.

JENSEN, M. C., AND K. J. MURPHY (1990): "Performance Pay and Top-Management Incentives," *Journal of Political Economy*, 98(2), 225–264.

KANDEL, E., AND E. LAZEAR (1992): "Peer Pressure and Partnerships," *Journal of Political Economy*, 100(4), 801–817.

KANDORI, M. (1992): "Social Norms and Community Enforcement," *Review of Economic Studies*, 59, 63–80.

KERR, S. (1975): "On the Folly of Rewarding A While Hoping for B," *Academy of Management Journal*, 18, 769–783.

KNIGHT, F. (1921): *Risk, Uncertainty and Profit.* Houghton Mifflin, Boston, Massechuchetts, Reprint, London School of Economics, London, 1946.

KRÄKEL, M. (1999): *Organisation und Management.* Mohr Siebeck, Tübingen, Chapter B: Hidden-action im LEN-Modell, pp. 62–77.

KREPS, D. M., P. MILGROM, J. ROBERTS, AND R. WILSON (1982): "Rational Cooperation in the Finitely Repeated Prisoners Dilemma," *Journal of Economic Theory*, 27(2), 245–252.

LAZEAR, E. (1979): "Why is There Mandatory Retirement?," *Journal of Political Economy*, 87(6), 1261–1285.

——— (1986): "Salaries and Piece Rates," *Journal of Business*, 59(3), 405–431.

——— (1989): "Pay Equality and Industrial Politics," *Journal of Political Economy*, 97(3), 561–580.

——— (1995): *Personnel Economics.* MIT Press, Cambridge, Massachusetts.

LAZEAR, E. P. (2000): "Performance Pay and Productivity," *American Economic Review*, 90(5), 1346–1361.

LEVIN, J. (1999): "Relational Incentive Contracts," revised chapter of MIT PhD Thesis.

MA, C.-T. A. (1994): "Renegotiation and Optimality in Agency Contracts," *Review of Economic Studies*, 61, 109–129.

MACHO-STADLER, I., AND D. PEREZ-CASTRILLO (1997): *An Introduction to the Economics of Information: Incentives and Contracts.* Oxford University Press, Oxford.

MACLEOD, W. B., AND J. M. MALCOMSON (1989): "Implicit Contracts, Incentive Compatibility, and Involuntary Unemployment," *Econometrica*, 57(2), 227–480.

MACLEOD, W. B., AND D. PARENT (1998): "Job Characteristics and the Form of Compensation," Discussion paper, CIRANO Document de travail.

——— (1999): "Job Characteristics, Wages, and the Employment Contract," *Federal Reserve Bank of St. Louis Review*, 81(3), 13–27.

MALCOMSON, J. M. (1999): "New Developments in the Study of Contracts on Labor Markets," in *Handbook of Labor Economics*, ed. by O. Ashenfelter, and D. Card, vol. 3B. Elsevier, Amsterdam, 5 edn.

MCAFEE, R. P., AND J. MCMILLAN (1991): "Optimal Contracts for Teams," *International Economic Review*, 32(3), 561–577.

MIRRLEES, J. A. (1999): "The Theory of Moral Hazard and Unobservable Behavior (Part 1)," *Review of Economic Studies*, 66, 3–21, reprint of a manuscript from 1975.

MOOD, A. M., F. A. GRAYBILL, AND D. C. BOES (1974): *Introduction to the Theory of Statistics.* McGraw-Hill Book Company, Singapore, 3 edn.

MOULTON, B. (1986): "Random Group Effects and the Precision of Regression Estimates," *Journal of Econometrics*, 32(3), 385–397.

——— (1990): "An Illustration of a Pitfall in Estimating the Effects of Aggregate Variables on Micro Units," *Review of Economics and Statistics*, 72(2), 334–338.

MÜLLER, A., AND D. STOYAN (2002): *Comparison Methods for Stochastic Models and Risks.* John Wiley and Sons, Chichester.

MURPHY, K. J. (1999): "Executive Compensation," in *Handbook of Labor Economics*, ed. by O. Ashenfelter, and D. Card, vol. 3. Elsevier Science B.V., Amsterdam.

NALEBUFF, B., AND J. STIGLITZ (1983): "Prizes and Incentives: Towards a General Theory of Compensation and Competition," *Bell Journal of Economics*, 14, 21–42.

NAPEL, S. (2002): *Bilateral Bargaining.* Springer, Berlin.

NELSON, F. D. (1977): "Censored Regression Models with Unobserved Stochastic Censoring Thresholds," *Journal of Econometrics*, 6(3), 309–328.

NEWEY, W. K., AND D. MCFADDEN (1996): "Large Sample Estimation and Hypothesis Testing," in *Handbook of Econometrics*, ed. by Z. Griliches, and M. D. Intriligator, vol. 4, pp. 2111–2241. North-Holland.

PAARSCH, H., AND B. SHEARER (1999): "The Response of Worker Effort to Piece Rates: Evidence from the British Columbia Tree-Planting Industry," *Journal of Human Resources*, 34(4), 643–667.

PANNENBERG, M., AND M. SCHRÖDER (2000): "Betriebliche Sondervergütungen in Deutschland," Discussion Paper 35/00, DIW.

PHELPS, E. S. (1973): "The Statistical Theory of Racism and Sexism," *American Economic Review*, 62(4), 659–661.

PRENDERGAST, C. (1999): "The Provision of Incentives in Firms," *Journal of Economic Literature*, XXXII, 7–63.

——— (2000): "The Limits of Bureaucratic Efficiency," distributed at PIPPS conference in Bristol, March 2001.

ROSETT, R. N., AND F. D. NELSON (1975): "Estimation of the Two-Limit Probit Regression Model," *Econometrica*, 43, 141–146.

ROTHSCHILD, M., AND J. STIGLITZ (1976): "Equilibrium in Competitive Insurance Markets: An Essay in the Economics of Imperfect Information," *Quarterly Journal of Economics*, 87, 355–374.

SALANIÉ, B. (1998): *Economics of contracts.* MIT Press, Massachussets.

SCHNEDLER, W. (2000): "Who Gets the Reward? An Empirical Exploration of Bonus Pay and Task Characteristics," Discussion Paper 235, IZA.

——— (2001): "Kooperation, Verantwortung und Leistungsentlohnung: Eine empirische Untersuchung," in *Entlohnung, Arbeitsorganisation und personalpolitische Regulierung*, ed. by U. Backes-Gellner, M. Kräkel, D. Sadowski, and J. Mure, pp. 39–50. Rainer Hampp Verlag, München und Mering.

——— (2002): "The Virtue of Being Underestimated: A Note on Discriminatory Contracts in Hidden Information Models," *Economics Letters*, 75(2), 171–178.

——— (2003a): "Informationsverwendung in Arbeitsverträgen bei Multi-Tasking," *Zeitschrift für Betriebswirtschaft*, 03(EH 4), 1–15.

——— (2003b): "The Prudence of Rewarding for A While Hoping for B," Discussion Paper 03/067, CMPO.

SCHNEDLER, W., AND U. SUNDE (2002): "The Hold-Down Problem and the Boundaries of the Firm: Lessons from a Hidden Action Model with Endogenous Outside Option," Discussion paper 464, IZA.

SCHWEIZER, U. (1999): *Ökonomische Theorie der Verträge.* Mohr (Siebeck), Tübingen.

SHAPIRO, C., AND J. E. STIGLITZ (1984): "Equilibrium Unemployment as a Worker Discipline Device," *American Economic Review*, 74, 433–444.

——— (1985): "Can Unemployment Be Involuntary? Reply," *American Economic Review*, 75(5), 1215–1217.

SHAVELL, S. (1979): "Risk-Sharing and Incentives in the Principal-Agent Relationship," *Bell Journal of Economics*, 10, 55–73.

SHEARER, B. (2000): "Piece Rates, Fixed Wages and Incentives: Evidence from a Field Experiment," University of Laval.

SHEARER, B., AND H. PAARSCH (2000): "Piece Rates, Fixed Wages, and Incentive Effects: Statistical Evidence from Payroll Records," *International Economic Review*, 41(1), 59–92.

SLADE, M. E. (1996): "Multitask Agency and Contract Choice: An Empirical Exploration," *International Economic Review*, 37(2), 465–486.

SPENCE, M. (1973): "Job Market Signalling," *Quarterly Journal of Economics*, 87(3), 355–374.

SPREMANN, K. (1987): "Agent and Principal," in *Agency Theory Information and Incentives*, ed. by G. Bamberg, and K. Spremann, pp. 3–37. Springer, New York, Berlin, London, and Tokyo.

TIROLE, J. (1999): "Incomplete Contracts: Where Do We Stand?," *Econometrica*, 67(4), 741–781.

TOBIN, J. (1958): "Estimation of Relationships for Limited Dependent Variables," *Econometrica*, 26, 24–36.

VANDERVEEN, T. D. (1995): "Optimal Contracts for Teams: A Note," *International Economic Review*, 36(4), 1051–1056.

WALRAS, L. (1874): *Élements d'économie politique pure.* Corbaz, Lausanne.

WILSON, C. (1977): "A Model of Insurance Markets with Incomplete Information," *Journal of Economic Theory*, 16, 167–207.

List of Figures

List of Tables

List of Symbols

E^F	feasible set	30
e^{FB}	first-best effort vector	26
E^{FB}	set of first-best efforts	26
e^{BSB}	balanced second-best effort vector	35
E^{BSB}	set of balanced second-best efforts	35
e^{BC}	balanced certainty effort vector	40
E^{BC}	set of balanced certainty efforts	40
e^{SB}	second-best effort vector	28
E^{SB}	set of second-best efforts	28
e^{C}	certainty effort vector	32
E^{C}	set of certainty efforts	32
$f(y, \theta)$	density function of y, parameter θ	104
$f^s(\nu^s y, \theta)$	conditional density given state s	106
$\tilde{f}^s(\nu^s y, \theta)$	contribution function for state s	106
$f_j(y\|\cdot)$	density function of Y_j	109
$F_j(y\|\cdot)$	cumulative density function of Y_j	109
$\phi(y\|\mu, \sigma^2)$	normal density; mean μ and variance σ^2	110
$\Phi(y\|\mu, \sigma^2)$	normal cumulative distribution function	110
$g(\beta, X^i)$	function linking expected bonus pay to X^i	134
ι	income of the agent	18
$\mathbb{1}_A(a)$	indicator function; $\mathbb{1}_A(a) = 1$ if $a \in A$ and zero else	104
J	Jacobian matrix evaluated at θ_0	131
$J(\theta)$	Jacobian matrix	130
κ	costs of benefit signal	72
λ	share of work-life under probation	73
$l(s)$	number of visible components in state s	106
$\mathrm{L}(\theta, \hat{\theta})$	quadratic loss when estimating θ by $\hat{\theta}$	63
m	number of signals in LEN-model	45
$\mathrm{MSE}(\hat{\theta})$	mean-squared error or estimator $\hat{\theta}$	63
∇_θ	first derivative with respect to θ	130
$\nabla_{\theta\theta}$	second derivative with respect to θ	130
$\Psi(e)$	surplus	26
Ψ^{FB}	first-best surplus	26
Ψ^{BSB}	undistorted second-best surplus	36
$\overline{\Psi^{\mathrm{SB}}}$	upper bound for second-best surplus	28
Ψ^{SB}	expected second-best surplus	28
Ψ^{C}	certainty surplus	32
Ψ^{BC}	balanced certainty surplus	40
$P(A)$	probability of the event A	105
$Q_0(\theta)$	limit of the objective function	127
$Q_n(\theta)$	objective function based on n observations	127

$Q^i_{\mathbb{W}}$	second derivative of objective function for observation i .	132
r	Arrow-Pratt measure of risk-aversion	45
ρ	signal-noise ratio	45
$\mathbb{R}$	set of real numbers	25
$\mathbb{R}^n$	set of n-dimensional real vectors	25
σ^2	variance	49
s	realised state of visibility	105
S	random state of visibility	105
$S(e)$	signal vector depending on effort	19
τ	transfer	26
θ	parameter vector	104
θ_0	data generating parameter vector	127
θ_j	uncensored expected value for Y_j	108
θ^i_j	uncensored expected value for Y_j and observation i	108
$\hat{\theta}$	estimator for θ	63
$\hat{\theta}_n$	estimator for θ based on n observations	127
Υ	incentive scheme	27
Υ^{SB}	optimal incentive scheme	28
$\Upsilon_e(\cdot)$	least cost incentive scheme to implement e	30
$u(\cdot)$	utility of the agent	26
$\mathcal{V}_j$	visibility set	104
$\bar{\mathcal{V}}_j$	complement to the visibility set	104
ν_j	visibility operator	106
$\bar{\nu}_j$	operator extracting unobservable components	106
v_j	realised visibility of Y_j	104
v^s	visibility of state s	104
V_j	random visibility of Y_j	104
$V(S)$	value of signal S	42
$\mathrm{VAR}(\hat{\theta})$	variance of the estimator $\hat{\theta}$	64
ω	guaranteed minimal wage	72
w	wage vector, $w = (w_0, w_1, \ldots)$	54
w_0	base wage	46
w_j	performance wage rate for j-th signal	46
X	task descriptors and control variables	108
X^i	task descriptors and control variables for i-th individual	134
x^i	realisation of X^i	108
y	realisation of a continuous random vector	104
Y	continuously distributed random vector	104
y_j	realisation of j-th component of the random vector	104
Y_j	j-th component of the random vector	104
$y^i_{(j)}$	j-th largest bonus component for observation i	108
z	real number, $z \in \mathbb{R}$	47

Index